DIARY OF A

PINT-SIZED

FARMER

A Year of Keeping Sheep,
Raising Kids, and Staying Sane

DIARY OF A
PINT-SIZED
FARMER

SALLY URWIN

DRG

David R. Godine, Publisher
Boston

Published in 2020 by
David R. Godine, Publisher
Boston, Massachusetts
www.godine.com

First published in Great Britian in 2019 as *A Farmer's Diary: A Year at High House Farm* by Profile Books Ltd

LIBRARY OF CONGRESS CATALOGING-IN-PUBLICATION DATA
Names: Urwin, Sally, author.
Title: Diary of a pint-sized farmer : a year of keeping sheep, raising kids, and staying sane / Sally Urwin.
Identifiers: LCCN 2020019000
ISBN 9781567926781 (hardcover)
ISBN 9781567926873 (ebook)
Subjects: LCSH: Urwin, Sally—Diaries. | Sheep ranchers—England—Diaries. | Sheep farming—England. | Family farms—England. | Farm life—England. | Country life—England.
Classification: LCC SF375.32.U79 A3 2020 | DDC 636.30942—dc23
LC record available at https://lccn.loc.gov/2020019000

Frontispiece: Ben, Sally, and, Lucy © Ian Wylie
End picture: Ben, Sally, Steve, Lucy, and Candy the enormously fat pony © Paul Norris

FIRST PRINTING, 2020
Printed in the United States of America

CONTENTS

Prologue

On a sunny spring March day, there is no better place to be than flat out in the straw of the lambing shed. The sun was streaming through the big double doors, and I decided that lying on the clean, dry bedding was a lovely place to have a snooze. Especially as I was surrounded by a flock of heavily pregnant ewes, who were calmly chewing their cud or sleeping stretched out in the straw. No one was due to lamb for a few days, so we all lay together, shifting a leg occasionally to get more comfortable, napping in the bright sunshine and storing up some sleep before lambing started ...

A party of visitors appeared at the lambing shed door and peered over the top to look at the sheep.

I hauled myself upright and staggered over to say hello.

They asked who the farmer was.

'I am,' I said.

They looked unconvinced. 'But ... who actually does the farming?' they asked.

'I do. Me. That's what I do. By myself. On my own,' I replied.

They looked around as if expecting my husband, wearing bib and braces, to pop up from behind a hay bale.

I tried again. 'My husband Steve, who owns the farm, is at his other work today. So, I'm actually in charge.'

I realised I was trying to persuade people I'd never met, who didn't know me, or even actually care, that I'm capable of looking after the sheep by myself.

They continued to look amazed, and after introducing them to a few of my (very fat and lazy) ewes, they walked back to the car park, making the odd 'Well, I never!' sort of noises.

Maybe I don't look like a farmer? I'm only 4 foot 10 (on a good day). Perhaps they expect all farmers to be big beefy men with ruddy cheeks and hands like spades.

I refuse to say, 'Oh, I'm the farmer's wife', as it makes me sound like I'm in a nursery rhyme or I stand in the kitchen making Yorkshire puddings and pots of tea.

Perhaps I should wear a name label, or a boiler suit with tractor logos all over it (if I could find one that didn't need two feet chopped off the ankle).

I never saw myself living on a farm when I was growing up. Myself, Mum, Dad and my older brother lived in a tall Victorian terraced house, right next to the North Sea, so my childhood was spent rock pooling, or on the beach, or going for long blustery walks along the seafront. My brother is a few years older than me, so I passed a lot of time playing on my own in our garden, most of the time pretending that I was in Enid Blyton's *Faraway Tree* or that I was riding my own horse. I was happiest reading a book, or playing with my stable of Sindy horses or trotting and cantering around the garden, stick in hand, deep in an imaginative game of showjumping at Hickstead.

I was good at English and History at school, and after A-Levels I did the expected thing and went straight to university. After my degree, I found myself a nice, safe, immensely boring office job. I was hired to provide 'marketing services' to a group of insolvency specialists at a huge company in the centre of Newcastle. It was as depressing as it sounds.

I wore smart suits and big heels, had 1990s blonde-streaked hair and got my acrylic nails refilled every four weeks. My main role was trying to think up marketing slogans to promote the company to businesses that were about to go bankrupt. The one highlight was Fridays, where everyone in the department used to pop across the road to the local wine bar for a long, boozy lunch before we all staggered back at 6 p.m., for ten minutes' coffee-drinking before clocking off for the weekend. Rumour had it that my department also had an account at the lap-dancing bar next door.

Working in an office was a huge slog: the humdrum 8 a.m. to 6 p.m., office gossip and politics, the fact that in the winter I'd get up in the dark and spend eight hours under flickering artificial light and then leave in the dark again in the evening.

The trouble was that I wasn't sure what I wanted. I didn't enjoy working in an office, but I couldn't see myself doing anything else. I had very little self-confidence and started to compare myself to my friends, and their burgeoning careers and increasing pay packet. But I was too shy and uncertain and didn't have the ambition to put in the long hours and endless meetings that a corporate career required.

I was also single, after a few relationships that had fizzled out. Everything felt dull and drab and boring and I felt rudderless and full of self-doubt.

Internet dating was in its infancy in the early 2000s, but I decided that I wasn't going to meet anyone on my wavelength in the bars and pubs of Newcastle. Instead, I signed up to a dating site called DatingDirect. I remember being very specific about who I wanted to meet, and searched for men who lived in the countryside, from Northumberland, didn't smoke, and were under 5 foot 8.

Trawling through the results, I spotted a picture of Steve. He was wearing the most hideous woolly roll-neck and smiling out uncertainly at the camera.

Our first date was in the middle of lambing, and I remember trying to pick up two newborn lambs, slippery and steaming, out of a freezing paddock in the teeth of a north-easterly gale. Over the next few weeks, Steve and I spent our time huddling in the lambing shed, mucking out pens, grabbing the odd takeaway and pint before rushing back to the farm to check the stock. I was very happy. I loved it all. Steve was just what I wanted: uncomplicated and straightforward and deeply connected to his farm and to the countryside.

I handed in my notice and gave away my suits; a year later, in 2005, we were married. I moved onto his farm, bringing as a slightly horrifying wedding dowry Cyril the elderly black-and-white cat and a small brown grumpy Shetland pony called Gladys that I'd bought off the side of the A1 in a fit of misplaced pity.

We all settled in beautifully. Cyril became a proper farm cat, skulking around on hay bales catching mice, and Gladys sank gratefully into life as a small, round 'field ornament' (as Steve calls our less useful animals).*

I loved walking around the farm, exploring the 200 acres of grassland, woods, crop fields and old farm buildings.

High House Farm is a small patchwork of beautiful rolling countryside, flattish clay fields, a damp ten-acre wood and two little streams called Sparrow's Letch and the Welton Burn. The farm itself is made up of a jumble of nineteenth-century buildings that are built solidly in dark cream and grey sandstone that glows golden in the sunshine. We have some new buildings as well: a modern lambing shed that we call the 'top hemmel'** plus a rather decrepit open-sided concrete-and-metal barn that is usually full to the rafters with hay and straw bales. One end of the farm is only 400 yards from Hadrian's Wall, and on a clear day we can see right across the beautiful Tyne valley.

After we got married I threw myself into the business of High House Farm. Encouraged by the government's shout to farmers that they had to 'diversify or die!', Steve took out a mortgage and converted the beautiful old granary buildings

---

* In my experience most farms usually have a variety of decrepit unsaleable animals hidden round the back of the sheds that have been kept on out of pity and become 'pets'. We usually have an assortment that has included Blind Sheep, Scabby Ewe and of course the Fat Pony.
** A 'hemmel' is a Northumbrian word for a shed.

into a microbrewery. In 2006, we took out an even bigger loan to refurbish the open-sided hay barn into a wedding venue and the old farm buildings into a tearoom and restaurant.

We had two children, Lucy and Ben, in 2007 and 2010. Those early years are a blur. I remember either being pregnant or with a small baby in my arms, running up and down the brewery steps, sorting out deliveries and trying to do brewery tours, while attempting to keep our young family warm, fed and watered.

As High House Farm Brewery grew, it became obvious that it was slipping away from us. We didn't have the time or skills to run the business and the farm as we wanted, and cutting corners and relying on agricultural contractors and temporary staff wasn't working. By a stroke of good luck, an employee said that she would buy the entire brewery, tearoom and wedding business, so we sold her the whole kit and caboodle and drew out a plan to rent out the farm buildings to her.

It was absolutely the best thing we could have done. Heather and her husband Gary took over the business, and through immense amounts of hard work turned it into a successful enterprise, holding weddings and receptions in our listed barns and brewing fabulous real ale. High House Farm Brewery is now fully booked with weddings for the next two years and the beer is sold as far afield as London.

Steve and I now work on the farm, alongside the brewery and wedding business. We have sheep and chickens and wheat, barley and oilseed rape crops. We also both work part-time to try and bring a little money in to pay the bills. We juggle our jobs, the farm and the kids. Money is tight, and our loans are big. But I'm just incredibly grateful that our family has the chance to farm in such a beautiful county, and to live the lifestyle that we've always wanted.

AUTUMN

Autumn is a season of preparation on the farm: it's the time we get the flock ready for the next year's lambing, buy new lambs to replace elderly ewes, sell our 'fat lambs' at market, plough and drill our crops for next spring ... and repair endless fences ready for winter.

Sunday, 3rd September

I'm half asleep in our double bed, enjoying my first lie-in for a while. I can just hear the low purr of the quad bike as Steve drives round the back field, doing an early morning check on the ewes. He'll have the kids and Mavis the collie sitting on the back of the bike, ready to jump off and start gathering the sheep together.

The curtains are drawn back, and I can see the swallows swooping past the window, gathering in bunches on the phone lines and chirruping loudly. They're getting ready to make their long flight to the African sunshine. One day soon we'll wake up and most of them will have gone and autumn will really have started.

All is peace and quiet. The cat is snoozing on my knees.

Suddenly the landline rings shrilly and I shoot up in alarm, knocking the cat to the floor. One of our neighbours is on the line: 'All your sheep are in the garden! They're on the lawn, and I have to say, they're making a bit of a mess.'

Oh Christ.

This is the second time this neighbour has found our sheep in their front garden. Last time the ewes were staring through their front windows trying to watch the telly.

I pull on yesterday's clothes and stagger out the front door.

Opening the neighbour's garden gate, I can see two white woolly bottoms right in the middle of their circular lawn. The grass is normally like a billiard table, but today it's covered with tiny hoof prints and black sheep droppings.

I get a bit closer and realise that 'all of the sheep' is just Button and her sidekick Keith. Button and Keith are seven-month-old pet lambs. 'Pet lambs' are the sheep that we bottle feed each year, due to their mother's rejecting them or not having enough milk. Keith is a chunky lamb with strong legs, a tight-curled coat and a fat bobble tail. He's a bit dim. Today he has rust-coloured streaks down his leg and back.

Button is a complete pain in the arse.

She was the third lamb out of triplets. Born tail first, she took a while to take her first breath. She's never really grown properly, even after we started feeding her on the bottle. She's tiny and has a strange shape, with a prominent spine and a sagging stomach. Her body reminds me very much of a woolly handbag. Her fleece is a lovely close-curled creamy white, which shades to a delicate chocolate brown on her legs and stomach. She has a neat black nose and big limpid eyes fringed with the longest eyelashes I've ever seen. She flutters these at Keith and anyone who might be holding any lamb feed.

Pet lambs tend to have no fear of people or dogs and Button is no exception. She sees gates and fences as challenging obstacles on her constant search for interesting feed choices.

She's started rolling under the bottom rung of the front gate and trotting up the driveway, nibbling on whatever garden plants catch her fancy.

This time Keith has tried to keep up with his girlfriend and has squeezed his roly-poly podge under the gate – hence the muddy, rusty marks.

Button looks round when I come through the gate and bounces over to start snuffling in my coat, looking for lamb feed, while Keith keeps stuffing lawn grass into his face.

Shooing them quietly, I manage to get the pair out of the garden and close the gate on their backsides. I hoist Keith up over the gate. He looks bewildered to be suddenly upside down with his four feet in the air, and struggles until I set him down on the other side. He's bloody heavy. Button tries to make a break for it up the drive, but I grab her by the fleece and push her under the gate.

She glares at me on the other side of the fence. Keith is two inches behind her.

I grab a couple of wooden sheep hurdles and wedge them in front of the gate. Their bottom rungs are lower, and hopefully will stop Button squeezing back under. Unless she works out how to nudge them open. I wouldn't put it past her.

Walking back to the house, I avert my eyes from the mess on the lawn and try to scuff a few pellets of lamb poo away from the drive with my feet. Back in the house, the kids and Steve are tucking into pancakes. It's just turned 9 a.m. – time to check round the rest of the flock.

Thursday, 7th September

It's a gorgeous autumn day. The leaves are just beginning to turn, and it's still warm, so I head down to the bottom of the back field to do a bit of long overdue walling.

Like most Northumbrian farms, our fields are hemmed in by a hodgepodge of grey stone walls and wooden fences, and it's a constant battle to keep them upright and patch any gaps before they get big enough for sheep to get through.

Normally (and if we have the money) we ask David,* a stonemason, who can rebuild a stone wall faster than anyone else I've ever seen. He's at least six foot tall with a shock of gingery red hair, and to watch him wall is to see an expert at work. He charges by the metre, so instead of paying him to patch the smaller gaps, I've decided to try to fix a stretch of wall myself, just to save a little bit of cash.

I don't have a lot of experience or skill but I start off with enthusiasm and survey a broken-down section of wall at the bottom of the back sheep field.

Our walls are made up of irregular-shaped yellow sandstone and grey whinstone, with a core of smaller pebbles that holds it all secure. But a sheep has pushed through this section and all the stones are lying scattered willy-nilly on the turf.

Sorting through the pile of rubble, I find the big 'cope

---

* All names and identifying characteristics have been changed as a) I haven't asked people and b) I don't want any cross locals throwing stones at me when I pop into our nearby shop or pub.

stones' (ones that will cap off the top of the wall) and put them to one side. The 'thruff stones' (through stones) are even bigger. These are the whin slabs that sit across the entire wall, holding both sides together. I heave them into a pile and start sorting out all the tiny rocks and pebbles that make up the core.

I begin by wedging in the larger stones at the base of the wall and work my way up, stuffing the central gaps with smaller rocks. It's a hot and heavy job, and after about two hours of solid work, I step back and have a look. I've managed to build in half the gap, but it doesn't look quite right.

It has a distinct lean to the right, and when I experimentally wobble one of the larger base stones, the wall sways from one side to the other.

Looking at the stretch I've done, I reckon all it would take would be one curious ewe to push gently against it, and then the whole flock would be out.

Sod it. I sit down on the grass in a sweaty heap and tip my head back to stare up at the clear sky and enjoy the sunshine.

It's beautiful down here. There's a patch of trees on the right where a buzzard has nested for the past two years. I can see one of the birds now, lazily circling above me and making mournful 'keeeee' sounds. We've also got a pair down in the ten-acre wood, and I've seen them up close when they land by the side of the road. They're huge, with deep brown feathers, and chocolate brown and yellow speckled wings.

Looking up, I can see a dark speck against the green of the field, and as I watch, it turns into the shape of my dad. He's come to see what I'm doing. He used to be a director of a large consulting firm, and since he retired in a neighbouring village, he's made it his life's work to help me out and tell me what to do.

'That looks a bit wonky,' he says, surveying the leaning wall. 'What you need to do is–'

He starts telling me what I should be doing and demonstrating while lifting and repositioning some of the bigger stones in the gap. In a few minutes the wall looks much more secure and sheep-proof.

'That's better,' says Dad happily.

I've got a flask of tea and we sit down on the turf and companionably share a drink out of the plastic cup.

I love moments like these. The sky is blue and there's a warm wind. It's very peaceful at the bottom of the big field, and we can't hear any car noise, just the occasional baa from one of our sheep. It's a moment of much-needed quiet before the kids come home on the school bus.

Saturday, 9th September

I'm stressing about my lack of time and amount of farm paperwork I need to deal with this month. In mid-rant, I look over and realise that Steve is happily watching tractor videos on his mobile phone.

We are two very different people. But strangely complementary as well.

I'm constantly talking, inherently anxious, love meeting new people, and I'm at my happiest chatting loudly over dinner to family and friends.

Steve is a true introvert. He finds meeting new people hard work, and much prefers talking to individuals on his own wavelength about something he's interested in. Like tractors. He has always struggled talking to strangers and has told me that he finds reading faces or understanding the subtext behind a conversation very difficult.

He's a deep thinker, and when he does get anxious, it's about things that really matter, like our lack of money. He does things one at a time, beginning to end, properly and slowly, laying out all his tools first, collecting all the pieces and admiring them before starting the job. He's great with numbers, accounts and complicated diagrams.

I rush at stuff, wanting to get things over as soon as possible and paying no attention to the detail. My mind constantly flips from one new thing to the next. Steve often sits thinking about nothing at all. (People pay good money to learn mindfulness to calm down their anxious minds. Steve does it naturally.)

His favourite things to do: tinker with a tractor or quad bike; plough a field in a perfect straight line; eat a hot curry; and put together an intricate model or piece of machinery.

My favourite things to do: worry; sleep; stare at horses; read a book on my own with a cup of tea; and meet my friends to laugh over the latest piece of news and gossip.

Steve makes me laugh and calms me down when I get upset. He's my rock. My safe place. Do we argue? Of course, but not as much as you'd think, even though we're such different people.

Sunday, 10th September

September is all about preparing our flock of 200 ewes for next year's lambing.

Our sheep are a mixture of different lowland breeds (Texel, North of England Mule and Suffolk) that are all prized for their mothering ability and capability of producing two or more lambs.

Our tups* are pure Texel or Beltex, which are big, muscular animals that throw chunky lambs that can be sold for good prices for high-quality meat.

When I first met Steve, I was bowled over by the fact that one sheep could be 'applied' to another sheep and then after five months, two new sheep would be born. Not coming from a country background, it seemed so clever! And surely the way to make lots of money! As in most things, it's rather more

---

* 'Tups' is the word we use to mean 'rams' (i.e. boy sheep). 'Tupping' therefore means mating between a ewe and a ram. Many people also use the word 'working' if they're feeling slightly coy. As in, 'How's your new tup doing?' 'Aye, he's working well', which usually means he's mounting everything in sight with loads of enthusiastic grunting.

complicated than that, and there's lots of careful preparation before the ewes even get to see the tups.

Last week, Steve disappeared for a whole day into Hexham and wouldn't answer my phone calls. He came back from the Mart with a pair of cracking Beltex rams. One of the tups won second in the Mart 'top tup' competition, and he's a beauty with a short stocky body and loads of muscle. They were both around £600, which wipes out all our savings, so we'll be eating beans on toast until Christmas and dressing the kids in sheep feed bags.

I've always had the job of naming our tups. They all come with their own pedigree names, but I find the usual 'Boyo' and 'Buster' completely uninspiring. Instead I rename them with the most porn-style monikers I can think of, thereby hopefully inspiring them to great things when they meet our ewes. This time, however, I got carried away with the fact that we'd bought two competition rams, and named them Wilfrid and Cuthbert – posh Northumbrian names.

Steve pointed out that calling tups after celibate seventh-century Christian saints wasn't exactly going to inspire their most virile performance – and so the renamed Thrusty Clappernuts and Randy Jackhammer are currently munching grass out in the front paddock. Hopefully their new names will motivate them into great things come mating season.

So today we're making sure our boys are ready for the mating season, which means looking at their teeth and feeling their balls for bumps or growths.

The boys will stay in their own field until the 5th of November, when they join the ladies for the highlight of their year: the chance to mate with as many ewes as possible. This means that due to the ewe's gestation of 150 days, our lambs will be due on 2nd April if all goes well.

Each tup can serve around fifty to sixty ewes, so we start feeding them up to get them in the best condition. It's bloody hard work, and after a good season of tupping, the boys get the rest of the year off, to lie in a sunny paddock and do whatever the sheepy equivalent is of scratching their balls and playing *Grand Theft Auto* until it's tupping season again.

Wednesday, 13th September

Today is a good ploughing day. It's sunny and there's a slight breeze, which will help dry up any moisture in the ground so the soil will be crumbly and easy to work.

I don't plough as I can't reach the brakes on our tractor. Everything from the level of the plough to the height of the seat is set by computer from the tractor cab and there's a complicated sequence of actions to take every time you set off. I can just about manage to reach the accelerator by sitting on a couple of cushions and rolling up my jumper to put in the small of my back. The brake is a different matter; to reach it I have to slide forward so my chin is level with the steering wheel before I can bring the tractor to a juddering halt.

Today I climb up into the 'girlfriend' seat for a few turns around the field. The girlfriend seat is a fold-down spare stool that juts out next to the driver's seat. It's unpadded and lacks armrests and I find myself sitting at a constant angle, as the tractor has one wheel in the bottom of the furrow and the other on top of the unploughed earth. It is literally a pain in the back.

Even so, I enjoy staring out the cab window, the rich smell of the newly turned soil and the satisfaction of seeing the plough converting dusty yellow stubble into long strips of brown. I love watching the seagulls swooping down for the worms and beetles unearthed by the plough and the chance of spotting deer, hares and partridge in the field.

The sharply pointed plough share (or plough blade) digs eight inches into the soil, so that the whole layer of top earth is turned over, burying any weeds and exposing the chocolate-coloured soil to the air. It's important to watch the plough out of the rear window to make sure it doesn't hit any unearthed boulders and shatter. Our fields are littered with huge blue-stone rocks that work their way up from the bedrock into the top soil. The corners of the fields have piles of massive stones, most of them scored with deep scars from being struck by a plough many years ago. Some of them must have been there a hundred, or even two hundred years.

Steve derives huge satisfaction from ploughing in straight lines, entering a zen meditative state. However, I find sitting at a tilt and driving at a constant speed of 4 mph rather tedious after the first few minutes, so I hop down when we reach the top of the field nearest to our house. Once I'm out of the cab Steve chugs away down the furrow again. It will take him two days to plough this biggest field, and he won't stop except for a quick break for a sausage roll and a can of coke.

Friday, 15th September

Another ploughing day. I can see Steve from my kitchen window. He's sitting in the tractor cab with the back window open and I can just hear strains of 'Thunderstruck' by AC/DC wafting across the field. The cab is air-conditioned and even has a small fridge next to the seat. He's in the lap of luxury.

There's a cloud of seagulls following him, looking for unearthed worms, and occasionally I see the tractor shudder to a stop and Steve urgently jump down to dig out an overzealous seagull that has been flattened by the top curve of the ploughed earth. That evening, Steve dashes in to eat his tea (standing up) and then goes straight back out. When the weather is good, I might not see him for twenty-four hours. Looking out the window at 9 p.m. I can see the tractor lit up like a Christmas tree, all fourteen headlights blazing, as he trundles over the ground.

Saturday, 16th September

Dad and I are drinking coffee together and I'm having a moan at him. I'm feeling down as I've put on a lot of weight. Being very short means that every extra pound shows, and I'm bursting out of my farm jeans and leggings. When I'm stressed I can eat for England, and I've been worrying about the farm's poor financial situation and how we're going to manage over Christmas and into next year.

Dad is full of sympathy, and he decides that we should *both* go to the local slimming club, which is held in our local village hall.

Dad has never dieted in his life, managing to regulate his intake of pies and beer so that he has stayed the same weight throughout his 50s and 60s. But now that he's nudging 75 he could afford to lose a few pounds.

The club is held in the evening, so after a stodgy evening meal to keep us going, we trudge off into the gloom, neither of us really looking forward to the experience.

The first happy moment is spotting one of Dad's missing hearing aids lying in the gutter outside the hall.

'Ahha!' he says. 'I thought I'd lost that.'

Dad wears hearing aids in both ears and regularly loses one or both by setting them down somewhere and wandering off. Lucy has made him a 'Special Hearing Aid Case' from an old egg carton, but it doesn't seem to make much difference.

Before he'd gone for a haircut yesterday, outside the village hall (Dawn's Mobile Hairdressers – £4.00 for men's short back 'n' sides), he had taken them out and put them in his pocket. One must have fallen out into the gutter.

He rubs the hearing aid against his jumper and shoves it back in his ear. Unfortunately, the battery has died. Like a lot of hard of hearing folk, Dad tends to overcompensate and talk loudly. As he's grown older, his tact levels have decreased, which makes for some interesting experiences when he meets new people.

In the village hall, a very nice leader sits us down and takes us both through the dos and don'ts of the food plan. Dad is fascinated. It's never occurred to him how to count calories, and he listens carefully, making notes in his notebook, and asking lots of long-winded questions.

'So, if I go cycling and stop at the café,' he asks, 'can I have a scone with jam?'

The leader sighs and explains patiently that no matter how much exercise you do, it still doesn't mean that you can go mad in the bakery section of the local tearoom. Dad's face falls.

The meeting is endless. The leader goes through each member's weight loss or gain and we all clap like performing seals. After a bit my hands are red raw.

Dad enjoys the weigh-in very much. He carefully undoes his shoes, belt and watch and makes a tottering pile of loose change on a plastic chair.

Holding up his trousers in one hand he starts chatting and joking to the ladies in front of him in the weigh-in queue. They think he's great. I'm alternating between embarrassment and pride as he cracks a few jokes and everyone roars with laughter.

We stand behind a bloke dressed in a grubby Barbour and wellie boots. I'm wondering whether he's going to take them off for his weigh-in, when he suddenly spins round. 'Do you know what you really need to eat?' he whispers hoarsely.

We wordlessly shake our heads. We've just joined and know nothing.

'Tripe. And bananas. But mainly tripe. The proper white ribby stuff from the butchers, not the bloody awful plastic crap you get in the shops.'

His breath is so bad that I can feel my eyebrows scorching. Dad takes a wobbly step backwards, still clutching at his trousers with one hand.

Fortunately, the strange man spins round again to face the front and spends the rest of the session rocking backwards and forwards on his heels murmuring to himself.

Dad is weighed twice, as he can't quite believe how heavy he's become.

'I used to be seven and a half stone when I was 18. I was so thin I would stand sideways and no one knew where I was,' he says sadly.

I'm not surprised by my own weight one tiny bit, as I have many years' expertise in dieting, and like many women, know how much I weigh to the nearest ounce.

After the meeting, however, Dad is brimming with enthusiasm, and hasn't been put off one bit by Tripe Man. 'I'm definitely coming next week. That was brilliant. It all seems so easy,' he says, as I nod vaguely in response.

Tuesday, 19th September

We need to replace some of the older ewes this year, so clutching our chequebook we go off to Hexham Mart for the 'Breeding Sheep' sale.

I'm always on my best behaviour at sales and slightly terrified of doing the wrong thing. I'd never been anywhere near the Mart before I met Steve, and didn't even realise it existed. I always want to fit in, and not be outed as a townie interloper.

Hexham Mart is a squat grey building on the outskirts of the town, centred around four auction rings and a huge space filled with animal pens. It's a hive of rural activity, with a great café, Mart office, Hubbucks agricultural merchants, hairdresser, insurance brokers, land agents and the Northumberland county show head office.

The café does the best fish and chips in Northumberland. The fish is fresh, the chips are crisp, and you get your tea (milk already added) poured into mismatched cups from a huge teapot that sits behind the counter.

Today we run our eye down the list of sheep lots, and Steve picks out the ewes he wants to see. Buyers can inspect the sheep before they go in the ring, and we make our way down to the pens to cast an eye over the stock.

We can't afford many sheep today. Farming is going through a slump; we've already borrowed money for new farm equipment to replace some that desperately needed replacing, and we're still paying the mortgage for the farm diversification into the brewery. It's always a juggling act as, like most small farmers, our fields and stock aren't extensive enough to cover costs as well as bring in a living wage.

We're not convinced we can afford to buy any more sheep, but last week we crunched through the figures with Andrew, our friendly accountant, and he's confirmed that buying a few more breeding ewes does stack up financially in the long run (as long as none of them get ill, injure themselves or die, of course). After some undignified pleading with our bank manager, we've managed to borrow a bit more money.

We can just about afford twenty new ewes, which will replace some of the older girls in our flock, as well as adding a few more breeding sheep to the farm. If we buy younger ewe lambs we might even be able to buy a few more, if the price doesn't go too high.

Drew, the auctioneer, starts the bidding as the first lot of sheep run into the ring. I find it weirdly relaxing to listen to his voice. All the auctioneers rhythmically string their words together and it becomes soporific if you listen for a while.

'Seventytwoseventytwoseventy*two*,' chants Drew, 'comeon, somecannyyounglambshere, who'll give me seventythreeseventythreeseventy*three*?'

Mart auction ring number four is set like a stage, with bleachers stacked up around a small sawdust ring. Most of the bidders arrange themselves around this centre stage, with their arms resting on the top rail.

They all bid secretively, raising just a forefinger or making the tiniest downward motion with their catalogue. It's not the done thing to show other people that you're bidding by waving your catalogue or arms around like a madwoman.

When the ewe lambs we have our eye on come into the ring, Steve gets Drew's attention by slightly tipping his catalogue and then bids by raising his right forefinger a few millimetres off his knee. We seem to be bidding against one of the bigger landowners, who's standing by the ring. The price goes too high and Steve shakes his head at Drew, signalling that he doesn't want to bid any more.

The next sheep lot come in: twelve Mule ewe lambs. They're nice and chunky, with bright eyes and good feet with wonderful speckled legs. They'd be perfect for us. Fortunately, there's not much competition, and after some surreptitious bidding we win. Each sheep costs us around £80. We decide to stop there, as we're terrified we'll overreach our budget.

Steve wanders off to pay for his new sheep and to back the trailer into the Mart loading dock. I go and have a final cup of tea in the Mart café and watch all the farmers and visitors catching up on the gossip and exchanging snippets of news. I'm beginning to recognise a few faces. There's a huge variety of people, from elderly retired farmers in ancient tweeds to much younger farmers, most of whom are wearing waterproof trousers and rigger boots. There are a lot of women farmers as well, which is great to see, in green wellies and waterproof coats, and some in fancy gilets and overalls with the name of their farm or flock embroidered on the back. Scattered through the crowd are kids, babies being looked after by their dads for the day, plus a few older children who must have blagged time off school. Lucy begged to come with us this morning – she adores the Mart and the café, and loves to go with Steve wearing her matching Claas overalls. But we're mean and packed her off on the school bus.

I buy both kids a chocolate bar from the counter, pick up a *Hexham Courant* and wander outside to find Steve and our brand-new sheep.

Wednesday, 20th September

Heather from the brewery catches me walking down the drive.

'There's some lambs in the yard!' she pants. 'Lizzie and I have shooed them back into the field, but there's still one left.'

Bollocks.

Thanking her, I race over to the brewery. As I arrive, I can see Button happily admiring her reflection in the hubcaps of the cars in the car park. She looks pleased to see me and bounces over to investigate my pockets for lamb feed.

Some walkers have been through her field gate and moved the sheep hurdle, and haven't put it back. Button must have been watching and waiting for her chance to roll under the field gate again.

This time Keith isn't with her – he's probably too dim to pick his way through the hurdle assault course.

A few interested brewery customers are watching as I pick her up and walk with her up the yard. I carry her like a baby, with her head on one shoulder, her bum cradled by one arm and her little hooves flailing in the air.

'Isn't she cute!' one little boy shouts. He gets to pat her head, while she snuffles in his ears and round his coat hood.

I dump Button in the chicken paddock. It has nice tall stone walls and the gate is low to the ground. Button is too small to be put in lamb, and she's too tiny to make a 'fat' commercial lamb at the Mart and be sold for meat.

'See if you can get out of that you little bugger,' I tell her crossly.

I still give her a lamb feed nut, though. The thought crosses my mind that maybe I'm unintentionally *training* her to get out of fields, by giving her treats whenever I find her.

Oh well. When I walk away, Button is carefully investigating and snuffling round the outside of the chicken house.

Thursday, 21st September

It's weigh-in day at the local slimming club. Dad has lost four pounds. I have put *on* half a pound. Bloody hell.

'Don't worry,' says the nice leader, 'you'll just be bunged up a bit. One good poo, and it'll all be gone.'

Saturday, 23rd September

We're sowing our winter barley today. Arable farming is a specialised science and highly technical, especially to outsiders like myself. The main things I've learnt is that:

The Basic Payment Scheme subsidy is a rural grant that all farmers are given by the government each year. To be awarded the grant our fields need to grow three different types of crops. High House grows winter barley, oilseed rape, wheat and spring barley, so each field is in a four-year rotation that improves the health of the soil and land.

Our crops are one of the most valuable things we grow on the farm, which explains the immense amount of stress and strain when we're trying to get the bloody seed into the ground and we're battling against bad weather.

Harvest happens late summer/early autumn, and is my least favourite time of the year, as I hardly see Steve, and when I do he tends to march around the house shouting at the weather forecasts on the TV.

Winter barley, planted in the winter, is a slower-growing type suitable for animal feed, and is harvested the following summer. Spring barley is a quicker-growing type and planted in the spring, and is usually sent for malting for beer. It is harvested at the same time as the winter barley, in the following summer/autumn. It literally took me about ten years to understand this. I think I'm naturally drawn more to the animals on the farm, and so the arable farming is a bit of a closed book.

I watch as Steve pours the winter barley seed into the drill, which is hooked up to the back of the tractor. Seed is expensive – like 'foreign holiday' sort of expensive – so he treats it like liquid gold, carefully tipping in each bag and shaking it out cautiously.

This winter barley seed variety is called 'Tower', and next September we'll harvest it to sell for animal feed.

I watch as Steve climbs into his cab, selects his Iron Maiden playlist for the day and rumbles out of the yard, hauling the seed drill behind him. It'll take him all day to sow or drill the seed, and then he'll need to flatten it into the ground with the two-ton roller that he pulls behind the tractor.

The tractor is the workhorse of the farm, and costs more than the average family house. It's Steve's pride and joy, and is kept in pristine condition, with tyres cleaned and paintwork washed and buffed each week.

Monday, 25th September

The vet came today to give Candy, our fat Shetland pony, her annual check-up and flu injection. When she saw Candy's enormous stomach and skinny legs waddling towards her out of the stable door, she did one of those reverse whistles that mechanics make when they see the state of my car.

'Are you sure she's not managed to get herself in foal?' she asked.

Nope. Unfortunately, it's pure lardiness from excess and illicit grass-eating.

Candy has put on a lot of weight, as she has (in the parlance of the horsey world) *no* respect for electric fences or anything else designed to help her restrict her food intake. She can work her way out of grazing muzzles and squirm under posts, and calmly bulldozes her way through electric fences. She probably doesn't even feel the sting of the electric shock due to the fat pads across her withers and enormous arse.

After a bit of poking and prodding, and a few more

involuntary noises at the size of her gut, the vet told us in no uncertain terms that the little Shetland had to be put on a strict diet.

So off I went to Carrs Billington farm shop in Hexham to buy the most electrically charged fencer unit I could find. I finally found an enormous German-made unit that promised 'Extra Power' with 'at least 0.22 joules shock power' that makes it 'an ideal partner for all your livestock'. Surely this would keep the fat pony in her field? I staggered out to Candy's field and started looping the unit over the fence before switching it on. It made a reassuring 'extra power'-type electric humming noise.

Candy looked at it suspiciously. She doesn't like anything new, so she retreated behind the water trough and inspected it from afar.

I was pleased with a good job done and prepared to triumphantly exit the field.

However, every time I attempted to close the gate, my arm and hand leapt upwards in an involuntary fascist salute.

I tried closing the gate for a second and then a third attempt, but each time I grabbed hold of the gate I found myself inadvertently making enthusiastic Nazi signals to my neighbours.

Finally, in confusion, I gripped the top bar of the gate tightly with my fists, and a huge zap shot up my arms and across my shoulders, making my heart do a double thump. I honestly thought I'd had some sort of stroke.

After a bit of a sit-down and a think, I realised that I'd accidentally hooked the electric fence over the metal gate sneck, so the whole gate was live, and I'd been unwittingly lighting myself up with 10,000 volts of top-quality German electricity.

I think Candy was laughing at me from behind the trough.

The upshot of the whole project is that the fat pony *did* attempt to walk through the electric fence last night, and after a zap and a squeal, now gives it a *very* wide berth.

All I've got to do now is make sure the huge battery pack stays charged.

Tuesday, 26th September

Today we need to check the tups again before they get let loose among their ladies.

Getting them in from the paddock is always exciting, as they lumber towards the sheep pens, lowering their head against Mavis the dog, who stays well out of their way.

Randy is one of the biggest tups I've ever seen. If he was so inclined, he could do some real damage with his hard, wedge-shaped head. He's also proven to be a grumpy bugger, and has chased me once or twice when I've been tipping sheep feed into his trough. I never turn my back towards him now just in case he catches me behind the knees.

Thrusty is gentler and tamer. He'll accept a sheep nut out of my hand, and once in the sheep pens he steps calmly into the tall metal walls of the sheep race, ignoring the testosterone-fuelled huffs and puffs of Randy behind him.

While the tups are in the medicated foot bath, Steve gives their teeth and testicles the once-over. I can see them all showing the whites of their eyes while Steve methodically goes along each tup's body, feeling for any lumps or bumps. Thrusty has a bit of a runny nose, so he gets an injection of Amoxicillin to clear up any infection. Apart from this, they all look well, and everyone seems to have a good set of gnashers.

The tups smell at this time of year – a very strong musky odour, which must be full of potent testosterone pheromones as all the ewes gravitate towards the gate to stare at them as they run back to their bachelor field. It must be frustrating for the boys as they can hear and see the ewes but can't get anywhere near them. Our flock of breeding ewes is getting a bit antsy too. As the days become shorter, they start to come into season, and they're flightier and more interested in whatever is happening outside the gate. Roll on tupping day.

Wednesday, 27th September

Steve is doing the farm VAT* return. He hates doing it, and before sitting down he's procrastinated much of the day, mowing the lawn, replacing a light bulb in the bathroom and clearing out his tractor cab. Everyone keeps out of his way when he finally fires up the accounting software. He needs to do three months' paperwork, and the misery of sorting through invoices and totting up how much we've paid out, compared to how much money has been trickling in, makes the whole house feel under a thundercloud. The kids are hiding upstairs and I'm sitting in bed with my iPad ignoring the swearing from downstairs.

Thursday, 28th September

Back to our slimming club today. Dad has lost another three pounds. He gets an award for it, a shiny purple one with a star down one side. I have lost half a pound, so I'm basically the same weight I was when I joined two weeks ago.

I sit in a childish huff all evening, sarcastically clapping all the skinnier people. Dad is very proud of his award and takes it home to put on the fridge.

I skulk home and open a packet of chocolate buttons. I don't want to go any more.

Friday, 29th September

Checking the sheep today I notice a big white lump at the far end of the field. It's absolutely bucketing down with great sheets of icy water, but peering through the gloom I can see that the lump is one of our ewes, lying motionless on her side.

---

* VAT stands for "value added tax," and since 1973 has been levied on most goods and services provided by registered businesses in the U.K.

I fire up the quad, and race towards her. I'm hoping she's not dead, as a) I'm very fond of our sheep, b) I'm not big enough to lift a dead sheep into the trailer and c) they tend to go off very quick and start to smell really bad.

When I get there I find poor Tilly the ewe stuck on her side in a puddle of water with a mass of blood where her left ear should be. She'd obviously laid down last night and rolled onto her back, and the sheer weight of the water in her wool had been too heavy for her to roll upright.

And then some scrotey creature (probably a badger or a fox) had crept up and gnawed off her entire ear while she lay there unprotected.

Poor soul. Unfortunately, this is fairly common – especially when the weather is bad, as the ewes' fleeces become saturated with water. Usually, the prone sheep get weaker and weaker, and then die. Crows were already circling, waiting to have a go at her eyes.

I hauled her by her sodden fleece and managed to get her onto her chest. She was shivering with the cold and her legs were so stiff that they stuck out behind her, so I wrapped my coat around her neck and massaged her legs to get them to bend underneath her backside. And there we sat for about thirty minutes in the driving rain, with me propping her up to stop her rolling back again and giving her a pep talk.

I don't have any proper waterproofs. Steve and I share the one waterproof coat we have (not because we're poverty-stricken, but just as I tend to wear fleeces unless it's really wet), and he had it on that morning. I was left with water trickling down the back of my neck and into my sleeves waiting for Tilly to try and get onto her feet. I sang a few songs to perk her up but every time she tried to get up she fell on her side again.

I call Steve, who is out buying sheep feed, and after about an hour, when we are both completely chilled through, he finally arrives back to help me manhandle her into the trailer and then into the warmth of the barn. She's had some strong antibiotics, had her wound dressed and been given some sheep nuts and hay. She'll be fine, and I got a cup of tea for a good rescue attempt.

We check our sheep twice a day, so thank god we caught her this morning – as I doubt she would have lasted another couple of hours.

Saturday, 30th September

Checking Button today, I notice that her spine is more prominent than it was, and she's looking a bit ribby.

She snuffles in my pockets for a few lamb nuts. Her eyes are still bright and she doesn't have the hunched back and lowered head of a poorly lamb, so maybe it was just the change of grass. After a chat and a neck scratch, she happily trots away to start grazing beside her ewe friends. If it keeps raining, we'll have to bring her in out of the cold, to see if that stops the weight loss.

When I leave the field, she's happily lying down, chewing her cud – her nose pointed skyward in typical 'relaxed lamb' pose.

Sunday, 1st October

Tilly One Ear is looking a lot better. Some sheep, when they've been attacked and lose an eye or an ear, go into shock and never recover.

We had one ewe last year who got stuck on her back and had her eyes pecked out. She died from septic shock later in the day. We've had lambs that have had tongues and eyes removed by crows. It's not Disneyland out here. Foxes eat chickens; ravens, crows and jackdaws attack newborn, sickly lambs or immobilised sheep; badgers and foxes will eat sheep alive, if they get the chance and the animal is too ill or injured to move. It's upsetting and dismal, but you can't blame them. Every animal is just trying to survive.

Tilly is up on her feet tucking into her silage. We'll keep

her in until the wound has completely healed. Sheep need a tag in each ear: one that displays the animal's individual identification number and one to show our flock mark. I'm not quite sure where I'm going to put the replacement tag.

Monday, 2nd October

This morning, I found a very smug Candy in next door's field, absolutely knee-deep in grass. She had spent the whole night busily ingesting 3 million calories in prime pasture. When I find her she's happily munching, completely unconcerned that the rich grass is making her fart noisily every time she moves.

She hasn't damaged the gate but inspecting the electric fence it looked like someone had deliberately unclipped the battery connectors.

Yesterday I noticed that the bride and groom from a wedding at the brewery had been in the horse field taking photos.* God knows why they went into that field, as it's a bit scrubby and is also dotted with huge piles of pony poo. When they were posing for their pictures, they were also standing right next to the run-off from the septic tank. The smell must have been distracting.

Maybe they unclipped the battery from the fence, so they didn't have to duck under the wire?

Whatever the reason for her escape, the outcome is always the same. A complaining Candy is put into her stable to deflate over the next twenty-four hours. She has straw to munch on, but spends most of the day hanging over her stable door trying to persuade brewery visitors that she's being mistreated and starved. It would work if she could stop farting and didn't look like a well-upholstered sofa.

---

* Wedding parties enjoy taking pictures with our farm as a backdrop. Which we don't mind at all! But of course, it's frustrating when people ignore our signs, leave gates open, and stray into areas that might be dangerous or easily damaged.

Thursday, 5th October

Most of the salesmen from agricultural supply merchants that visit us on the farm are lovely. And polite. And just want us to buy something from them. However, we have one geriatric feed salesman who is always turning up at the front door, when Steve isn't around, and demanding I give my husband feed supplement samples while staring at my boobs. Today, he turns up at the busiest time of the day, straight after school pick-up at 4 p.m., when I'm feeding children and animals.

'Is your husband in?' he asks. He won't talk about the farm to me directly, as I'm not a man, and therefore obviously know nothing about farming or sheep.

I'm wearing the unsexiest outfit in the world: a roll-neck jumper, leggings and slipper socks. It doesn't stop him though, and as usual, while he's talking, his gaze slips down to my admittedly magnificent bosom.

'No. No, he's not,' I answer distractedly, trying to stop the cat from running out the door onto the road.

'Could you give him this then, dear?' he replies, pushing a heavy sheep lick tub into my arms.

'No, I really fucking couldn't, you patronising old bugger.' I don't really say that. Instead, I smile weakly, take the lick, and promise to pass it on to Steve.

Friday, 6th October

Button isn't very well at all, and I can't work out why. She's looking very thin and ribby and today I've found her lying down in the gate entranceway. Her eye looks dull and she's obviously not eating anything as she's so skinny. I carry her into the shed and make up a little pen with some straw, feed and a tub of water.

Steve can't work out what's wrong either. Some lambs just don't thrive, and maybe her difficult birth has impacted on her digestion or the way she metabolises her food. We give her some antibiotics and painkillers in case there's an infection.

I sit with her for a bit, stroking her neck and rubbing her ears. I do this to myself each year in the months after lambing: get attached to some poor pathetic scrap of a lamb who becomes very friendly but then ultimately doesn't survive. Button has been so full of mischief though. It isn't fair.

'Could we give her a bottle of milk?' I ask Steve.

'You can't give a weaned lamb milk,' he replies gently. He knows better than to tick me off for getting attached. For all his big farmer grumbling, Steve is just as soft as I am. He pushes the water bucket closer to Button and splashes his fingers in it, but she turns her face away.

'Come on, leave her to it. She needs peace and quiet for a bit if she's going to get better,' he says. He's right. He grabs my hands and pulls me up from the straw and we walk back to the house together.

Saturday, 7th October

I'm spending a lot of time in Button's pen, trying to get her to eat or drink. She'll dip her nose into the water, but then turns her head and lies down on the straw. She's weak and listless and looks like she's given up.

We ring the vet for advice and try everything – heat lamps, vitamin injections, antibiotics and medicine. Getting the vet to come out to see Button isn't an option. It's a hard decision, but this tiny lamb is only worth around £60–£80 full-grown, so asking the vet to visit wipes out any profit at all. Steve has seen cases like this before. If the animal has some type of internal deformity or weakness, they're just not going to survive.

Sunday, 8th October

Button is still very poorly. Her eyes have sunk, and you can see her sides labouring as she's trying to breathe. Steve tells me we can't let her go on as she is, and that she needs to be put down to stop her suffering.

I beg him for one more day, and spend hours propping her up, holding the bucket to her nose, trying to get her to drink. She is just not interested but lies quietly in the straw. I leave her in the afternoon to make tea for the kids.

Later that night, after the children have been put to bed, Button dies. We find her little body stretched out in the straw. When I spot her, Steve takes me straight back to the house, tears pouring down my face. It's just one lamb, and I know I'm 'too soft', but I've become completely attached to the tiny animal. He goes back later to bag up her body and put it in a corner of the farm to wait for the knacker man. The kids are upset, but not as bad as I thought they would be. Maybe they've been inured to this – the sudden deaths and disappointments on a normal farm.

Monday, 9th October

I wake up with swollen eyes and a blotchy face. It's a gorgeous morning though – fresh and bright with a frost lying in the furrows of the front field. The ewes are cheering me up as well. They look fabulous at this time of year, with clean white fleeces and well-padded, chunky bodies.

Mabel is one of my favourite ewes. She's a sort of milky tea colour, with a tan fleece and a very handsome russet face. When she spots me leaning over her fence she huffs across the grass, her belly swinging from side to side, to demand a handful of sheep nuts. I keep a secret supply of them in my pocket, and distribute them indiscriminately to ewes, chickens, tups, lambs and the pony. It drives Steve mad.

'This is not a bloody petting zoo!' he shouts when he sees

me surrounded by a selection of farm animals, like a cut-price fairy-tale princess.

Mabel snuffles in my pockets for a few more nuts then, realising I haven't got any more, she submits to a neck scratch. I've realised that sheep aren't that fond of people patting their heads. I used to try and scratch our sheep between their ears, but they seem to prefer being scratched under their neck. Mabel eventually starts back across the field to the rest of the flock, nibbling here and there at a tasty blade of grass. Picking my way back across the garden, I go in to start making breakfast for the kids. I'm still upset about Button, but the day-to-day work on the farm distracts me from dismal thoughts, and all the other animals still need care and attention.

Thursday, 12th October

Tilly the one-eared ewe is looking a *lot* better. The antibiotics have kicked in, and she's bobbling around the shed with a couple of lamb friends, busily scrounging for overlooked sheep nuts in the straw.

We've eventually decided to put both tags through one ear, so now Tilly is sporting two dangly plastic earrings. We'll wait until her wound is fully healed and then she needs to go back out with the rest of the flock. She's very tame, and chases us around the sheep pens asking for food.

Instead of a normal 'bahhh' noise she makes a high-pitched 'creeeeeaaaak' noise. It sounds as if her vocal chords have been paralysed in some way, or she's suffering from a sore throat. Maybe the trauma of being stuck on her back has affected her voice.

Whatever's happened, it makes it very unnerving to go and check on her late at night. She likes to stand on her own at the back of the shed in the pitch dark and 'creaaakkkk' softly at me from out of the shadows. I'm hoping her proper voice might come back after she's had a bit more time in the warmth.

Saturday, 14th October

It's a cold autumn day with a smattering of frost on the rigg tops* in the back field. Mavis and I gather up the sheep for the morning check, our breath steaming out into the frigid air. I tell Mavis to lie down, and she obediently flops onto the grass, her long pink tongue lolling.

Checking over the sheep, I see some dark patches on their wool. While I watch I notice some of the lambs are scratching themselves against each other and the fence rails. Catching one with my shepherd's crook, I look through the wool, trying to find out what's making them all scratch. It can't be flies. It's almost winter, and far too cold for maggots.

The sheep's skin looks pink and raw, and wool has been rubbed away, leaving cracked, sticky skin. It's scab. This is a disaster. Scab is caused by tiny mites that burrow into an animal's skin and eventually causes sheep to stop eating, become poorly and lose most of their condition. Treating scab involves injecting each sheep with a remedy called Dectomax. It's easy to clear up, and the sheep will recover, but the withdrawal for Dectomax treatment is 100 days, which means we won't be able to take any lamb or sheep to the Mart until the 100 days is over, and that will be after Christmas.

Steve and I were relying on selling our lambs for meat and using the income to prop up our bank account. Now we can't.

I chew on my lip, deciding the best way to tell him. He won't be happy.

'What are we going to do? I've got bills coming out of my ears! I need that lamb money to pay for the last month!' he shouts when I tell him.

We make a cup of tea, and ring David the vet. He comes out straight away and inspects the sheep.

'I'm sorry. It is scab.' David knows very well how this

---

* Riggs are part of 'rigg and furrow', a system of ploughing used in the Middle Ages that produces distinct built-up ridges and dips or furrows in a field.

diagnosis will affect our bank account. He promises to leave out the medicine for us to pick up from the practice in Hexham.

With heavy hearts we start gathering in the sheep. They all need to be injected: tups, ewes, breeding lambs and fat lambs. They'll get better soon enough, and the scab will go, but we're devastated at losing the income at one of the most expensive times of year.

After discussing it, we decide to take all the lambs into the shed to keep them inside over the winter. It'll be more expensive, but at least we can keep an eye on them, fatten them as much as possible, and then sell as store sheep in the January sales.

Christmas is becoming a worry: how to buy presents, pay outstanding bills, buy gifts for the family and pay for heating oil. I take on more freelance marketing work – anything to get in a bit of money to pay off debts and keep our heads afloat.

Monday, 16th October

I'm mucking out the pony's stable of a week's worth of straw and dung. Candy stands in the corner of the barn munching on a hay net, watching me out of the corner of one eye.

Wherever you stand she always makes sure she can see you. She's a very cautious pony.

She spent a lot of her life in a busy riding school, teaching hundreds of small children how to ride. She hates anyone fussing around her head and ears, and I wonder how often she had children jabbing their fingers into her eyes or pulling on her mouth and ears. Because she's only small she has been ridden mostly by tiny beginners, with flapping legs and harsh hands.

Her passport says she's 18 years old, but the vet reckons she's nearer 26. Small ponies often live past 30, but Candy has arthritis, Cushing's disease and Chronic Obstructive Pulmonary Disease. This causes sore joints, excessively hairy coats and shortness of breath.

The vet reckons she's had a few foals, as her belly has sort of ... slipped, and she's distinctly baggy around her bottom.

She's very, *very* clever. On a lead rein she's a saint. In a riding school she will walk and trot like an angel, and even (very creakily) go over a small jump. In a field, with no one leading her, she's a holy terror. Lucy often rides her without a lead rein, and after about ten minutes Candy decides that everyone has had enough, and will grip the bit between her teeth, canter to the gate and give a good shake so that Lucy falls off.

Wednesday, 18th October

It's been raining now every day for a good fortnight, and the land is soaked. Eventually the rain clears, and the forecast shows three dry days in a row.

Steve races out to sow the field next door to the house with spring barley. With the tractor and a combination drill (which combines the ploughing and the sowing mechanism) he sows seeds into the soil and then flattens the field with the two-ton roller to squish in the stones and maintain the moisture in the soil.

He works until after midnight, but it means that our spring barley is finally planted and will be ready for harvest around next September. Eventually the barley crop will be sold to a grain buyer and be sent away to be malted and made into beer.

Friday, 20th October

I've not had a good day so far, so I'm sitting in the warm, wrapped in a blanket with the cat on my knee, avoiding everyone.

It started this morning.

My very posh neighbour, Sybil, rode past me on her hunter. I was trying very hard to get some lambs through the gate into the yard. She bellowed something like, 'I say, jolly good, get the bally things what jolly good what!' at me. Sybil has such a cut-glass accent that it's difficult to understand her. She's as posh as the Queen. She's probably related to the Queen. I just smiled and waved weakly, and then tripped over the dog.

Then when I walked past High House Farm Brewery, David, the brewery assistant, shouted over to me, 'Oi, Sal! Next time you're getting dressed in the morning, close your curtains, will you?' The horrors. He must have seen me in all my glory, boobs swinging and scratching my arse, while silhouetted against the morning sky. And the poor lad must only be in his mid-20s. I've probably scarred him for life. I'm never going to open the curtains again and have decided to spend my life swathed in a full-length floral housecoat.

And after last night's late check, I must have forgotten to lock the fat pony into her stable. I found her early this morning, down the road, in a neighbour's ornamental shrubbery, thoughtfully munching on their pampas grass. She'd spent a happy evening investigating all the old sheep feed bags in the shed, rubbing her arse on Steve's quad bike and decimating a nearby garden. Fortunately, nobody seemed to have noticed, so I hauled her back into her field and tried to ignore all the pony hoofprints across their immaculate lawn.

Saturday, 21st October

We have a sheep called Scabby Ewe.

She has an undershot jaw (called a 'shuttlegob' in the local dialect), meaning her bottom teeth don't properly meet her upper palate, so she finds it very difficult to chew hay or grass. She and her flock have been out in the field the whole summer, but Scabby can't chew the pasture well enough, so she's very, very skinny.

Inevitably, Ben has adopted her as 'his most favourite pet ever in the whole world'. When we tried to persuade him that one of the healthier, cuter sheep would make a much nicer pet, he was adamant that Scabby was his favourite ewe, even though she looks like a toast rack on legs.

Due to her jaw deformity, she's never going to be able to keep on enough weight like the other sheep. Some farmers would have her put down. But the kids have a minor meltdown when this is mentioned.

She did manage to raise one lamb this year. Just. It was a bit touch and go, but we carefully siphoned in lots of extra sheep nuts so that she was able to make enough milk for her single baby.

All farmers have the odd mangy sheep, and they normally put them in the back fields, well away from the road, so that no one can spot them and judge their stock-keeping. Poke around any farm and you'll eventually come across some cheerful, threadbare animal living the life of Riley in a tucked-away stable.

Because we're not heartless and miserable, we don't begrudge Scabby this extra feed. But we're going to have to make the decision what to do with her. We can't put her into lamb again, as she (and the lamb) may not survive.

Scabby's future lies in the balance. We can't even sell her 'fat' (for meat) as she isn't anything *near* the weight she should be, and would probably fetch around 20 pence, before everyone laughed us out of the Mart.

She's ridiculously tame and will happily follow you around the farmyard. Whatever I'm doing on the farm, Scabby is beside me, usually with her nose in my overall pocket, on the scrounge for extra food. She also has a rather nice fluffy bit between her ears where she enjoys a good head scratch.

So, I think she's destined to become a field ornament, just like Button the undersized lamb (R.I.P.) and last year's Blind Sheep,* which had some kind of brain problem and couldn't

---

* Blind Sheep used to graze quietly in the paddock, until he suddenly realised that he couldn't hear any of his sheepy friends. Then he would run urgently round and round in circles until he heard one of his flock and bobble unsteadily after them. At night-time he used to cuddle up next to an elderly ewe that didn't mind his heavy breathing and general dimness.

see anything. Eventually these elderly pets die of old age, and Steve breathes a sigh of relief, until the next lambing season brings along another few decrepit characters.

Sunday, 22nd October

Living on a farm isn't the solitary, hermit-like existence you might expect. It's not exactly Piccadilly Circus, but we do meet a fair few different people, mainly a result of the farm being a wedding venue, but also due to walkers, cyclists and riders wandering past our house.

I had a great chat with a man on a horse today. I saw him on a bay mare, leading a short, fat bay cob down our road. Being completely shame free, I stood on our garden gate and loomed at him over the top to ask where he was going, what he was doing, what his horses were called and where he lived.

When he got a word in edgeways he explained that he lived in Great Whittingham, was an engineer and liked to go on long hacks when he could, and that his ponies were called Lyn and Betsy. Then he added, apropos of absolutely nothing, 'Sometimes in the evening I like to cook naked in my back garden.'

'Gosh,' I said enthusiastically, 'that must be ... chilly. What happens if the neighbours catch sight of you?'

'Oh, I'm known for it,' he replied airily. 'I think they expect to see me out there, so they all look the other way.'

I wouldn't. If I lived next to him I'd be riveted. Not in a peeping Tom type way, but I'd want to know what he was cooking on his barbecue, and whether he wore an apron to protect his tender places from sparks, and whether he sat on the grass or ate his steaks standing up in a 'wild, hunting man' stance.

I invited him into the garden for a cup of tea, so he tied his horses up to the gate. Betsy did a wee on the drive, and I fed the man cake and asked him questions about his ponies and his hacking and living in the village. He said he'd knock on my door next time he came past. It's nice to make a new friend. Even one who is occasionally naked.

Monday, 23rd October

I'm furious today, as three of our five chickens have been snatched by a fox. They were stolen during the day, and after calling them for ages I found the inside of the chicken house was absolutely plastered in feathers and blood. Only foxes would take all three chickens in one go.

Steve was really upset – we'd raised Doris, Tabitha and Elsie from chicks, and they used to feed from our hands and sit on the tractor and join in people's weddings, even when they weren't invited. I'd spent many a happy hour avoiding the kids by sitting in the veg patch digging up worms for a grateful Tabitha.

I tried to be matter-of-fact with Ben and Lucy, but faced with a couple of wobbling lower lips had to have a rethink.

'Maybe ... maybe a mummy fox was so hungry and couldn't feed her three fox cubs, and had to take our chickens so that they wouldn't starve and they really didn't mind, as they were sort of pleased to give the hungry babies a good meal.'

It seemed to work. 'It's a farm,' said Lucy in an off-hand way, 'and things eat other things.'

Exactly.

I do feel sympathy for foxes, and I'm actually glad of the 'hunting with dogs' ban, especially since it seems to have attracted more people to the hunt than ever before, which can only be a good thing for employment in the countryside. But it's hard when you're faced with a sudden chicken massacre.

We've never had much luck with the chickens. Maud died by being suffocated one cold night when all the other chickens sat on her head. I don't understand this. If someone stepped on my head when I was asleep I'd be extremely bitey. Perhaps they all sat on her at once, and she was taken by surprise? Sounds rather premeditated.

We put Maud's body in the bin, which fascinated Ben, who then told the health visitor that 'Daddy put Maud in the rubbish', making it sound like we'd disposed of an elderly relative.

I clean out the chicken shed and give a pep talk to our remaining two chickens (Marjorie the Fourth and Ethel) about personal safety and self-defence. I've also put up more chicken wire around the perimeter of the shed, although it's beginning to look a bit like the film set of *Escape from Colditz*. The kids and I are keeping everything crossed that the fox (or her hypothetical cubs) doesn't come back.

Tuesday, 24th October

One of my most frustrating conversations happened last week, when I was bedding up pet lambs in the top shed. A winter wedding was in progress, and a wedding guest was floating around, watching me wrestle with the straw bale. When I wandered back out with the wheelbarrow he came up and said, 'Hi, I'm Martin and I'm a militant vegan.'

These were his exact words. He was wearing a tweed suit, had his hair in a pony tail and sported a rather fabulous ginger beard. We stared at each other for a bit, and then he showed me his plastic shoes. They were bright blue. I showed him my Ariat 'anti-fatigue' wellies, and how I could store my mobile phone in the flappy bit at the top.

'Why do you keep them, then?' he asked, gesturing to the pet lambs. They were all lying down in the clean straw, thoughtfully chewing their cud.

I suddenly wanted to grab him by the lapels and shout ranty, self-pitying, incoherent thoughts about how we keep them for meat, but you should see how well kept they are, and how we flog on looking after even the most pathetic scrap of baby lamb to make sure they're well fed and happy. If Steve and I were just out for profit, we'd skip all the care and good feed and tube feeding, and just knock the pet lambs on the head, and madly fatten up the rest and sell them as soon as they were heavy enough. But we don't. We care for the tiny buggers and we're proud of our farm assured status and we never have enough money as farming is going down the pan ...

I didn't say any of this but mumbled on a bit about our quality animal care and living conditions.

'Those lambs look all right,' he conceded, 'but I've seen videos online of some terrible places.'

'Well, we're not one of them,' I said stoutly, and he vaguely nodded his head in an unconvinced way and went back to the wedding.

I couldn't care less what people eat, as I'm usually too busy struggling with my own carb-heavy diet and ever-increasing waistband, but Martin's question took me by surprise and I felt all defensive and wanted to make him *see* what we're trying to do on our farm. I'm not convinced it worked though.

Wednesday, 25th October

Today, as Steve is away, I have to feed the outside ewes myself.

The sheep nuts for one evening feed come in a 25 kg bag. When Steve feeds the sheep outside, he opens one corner of the bag and, in one smooth motion, hoists it onto his shoulder and pours the feed straight into the trough, with the sheep trotting along behind him.

When I do it, I can't lift the bag, so I open one end and drag the bag behind me so that a bit of feed drops out and all the sheep gallop over to me to get it and I fall over. I'm then trampled by lots of hungry fat sheep, and Steve gets cross as all the feed is spread across the ground rather than in the trough.

Today I have a plan. I open the sheep nut bag on the quad bike, and then scoop out the feed with a shovel into each trough, shooing away any sheep that threaten to flatten me. It takes *ages* but at least I didn't fall over, no sheep nuts were wasted, everyone got fed, and the feed sack was empty.

Being under five foot gives me lots of farming problems – the main one being the fact that I have no leverage in wrestling a sheep onto her back to trim her feet. Steve says that tipping a sheep is 80 per cent technique and 20 per cent brute strength, but as I have very short arms like a *Tyrannosaurus rex* I can't reach to trim a ewe's rear feet when I'm holding her on her back. And lambing brings extra problems – keeping a ewe still and then reaching round to pull out a lamb is a physical impossibility with tiny, weedy arms.

I wonder if *Countryfile* would do a segment on height-disadvantaged farmers?

Thursday, 26th October

I met an old family friend today in the village. He knew both Steve and I before we were married. We exchanged a bit of chat about the family and then he said, completely out of the blue, 'I'm so glad that you two found each other.' I knew what he meant, but it did sound as if we were a couple of old wrinkly turnips that had been left on the shelf too long.

Friday, 27th October

I've been doing some useful work today. In the last rainstorm an old ash tree fell in the East Field. It must have been rotten, as the whole trunk split in two, leaving the canopy to crash down into the grass field, scattering dead branches all over the pasture. Broken up and dried out it'll make great firewood for the coming winter, so Steve trudges off to start chainsawing it into logs. I refuse to use the chainsaw. Cutting up a tree is a skilled job, as it's very easy to either chop into your own legs or trim the wrong bit and end up pinned beneath a large branch.

I stand and watch for a bit, as Steve methodically works through the ash branches. After a while I decide to go and start on one of my favourite jobs: clearing out a stream.

Sparrow's Letch runs through the bottom of the East Field. It's a bonny little watercourse that trickles between two deep banks. The water is cold and clear and runs through a jumble of deep pools and shallow waterfalls, over a higgledy-piggledy bed of stones, boulders and clay.

During the winter months the stream becomes a rushing torrent that carries tree branches, logs and old fence posts to end up jammed against the water gate, which is built across the stream to stop stock wandering between the two fields.

I mooch off to inspect the mound of branches and leaf litter that is piled up against the gate. If it wasn't cleared, the blocked gate would eventually completely dam the stream, which would then flood into the pasture, drowning the grass and stripping away the soil.

Pulling on some gloves, I start to haul away broken tree branches and fence posts to allow the water to run through the gate – so satisfying.

It takes a couple of hours to clear the trash, and when I've finished the stream is running completely clear beneath the water gate and I have a huge pile of old tree boughs, holly and hawthorn branches, split fence posts and leaf litter piled up on the stream bank. Mixed in with the organic stuff are plastic lick buckets, the odd empty sheep feed bag and half a drainage pipe, which has all been washed down by the stream.

I've discovered all sorts of interesting bits and pieces in the past while doing this job, including old green bottles, a large deer antler and two rusted milk churns.

Steve has finished chopping up the tree, so we load the logs into the trailer and take them back to stack in the farmyard. In a few weeks they'll be seasoned and we can split them into smaller logs for our fireplace.

Saturday, 28th October

It's been below freezing for a few nights in a row and that means there's been some gorgeous clear, cold and (above all) dry days.

The drier weather means that I've been pottering around the fields, admiring the frosty trees and holly bushes, like a short, tweedy nature sprite. Everyone (including me) has been doing some attempts at exercising. I've even made Candy trot up the 'long pull' to the wood. She hates it, so I lead her while jogging on the far side of the path; otherwise she tries to take a massive bite out of my shoulder. As a name, Candy really doesn't suit her. I always think Shetlands shouldn't have cutesy names. The ones I've known have been called things like Angel or Sweetie, which they most definitely weren't. I did see that the runner-up in the Shetland Grand National this year was called 'Idiot', which makes me very happy.

Today it's so bloody cold that I have developed chilblains. I think they must have been brought on by getting cold feet while gathering wood and feeding the sheep. So far, I've been told to wee on them, or rub them with an onion. Sometimes it feels like I'm living in the Middle Ages ...

Sunday, 29th October

Our two chickens are bored. Due to the avian flu threat, all poultry owners are supposed to keep their 'chucks' inside, so we've got ours penned up in a stable. They've got a roosting bar and a nesting box and everything, but every time I go and see them I'm treated to a long litany of clucky complaints which I imagine are about how bored they are, how it's so crowded and what Marjorie said to Ethel by the water bowl. I've started throwing sprouts in through the top door (they bop off their heads in an amusing way) and the chickens race after them, grabbing the sprouts and hurtling into a dark corner to gobble them down as fast as they can. Hopefully it won't be long before we can let them out again.

Monday, 30th October

We've sent our decrepit 'cast' ewes to the Cast Sheep Sale at the Mart. These are the ancient souls that have no teeth left, have non-working udders or, in one unfortunate case, have a huge misshapen hernia that bobbles alongside their body.

They all went under the hammer along with a selection of fat lambs. The best price for the old ewes was £72 per sheep.

As I'm writing this, Steve is in the background shouting, 'Don't tell them how much we got! Otherwise everyone will know!'

In my experience, farmers don't like giving away secrets. And they often take the 'more scenic route home' basically to drive round a bit and see what everyone else is doing. That's why we all put our best, youngest, prettiest sheep out in the front field and then hide the scabby, lame ones round the back. Steve usually gives me a running commentary as we drive past farms on the road to Hexham.

'Look at the ploughing on that. God, he's cocked it up round the electricity pole. Christ, look at the water on that wheat field! Now, what I would have done', etc.

I can't think of any other industry where people are able to stare over a hedge in order to freely comment on the way you do your job.

Tuesday, 31st October (Halloween)

A good (and true) ghost story for you. Since we renovated the old farm buildings back in 2003, and constructed the tearoom in 2013, some peculiar things have been going on.

Heather hears footsteps across the floorboards outside in the malt loft, when there's no one there. It's not the sound of an old building settling or creaking, but rather the clear 'slap, slap' of someone in leather shoes walking across the floors. We've all heard the noise of someone else running across the old granary, from one side to another. When you look out, the room is completely empty.

When Andrew the chef comes in to set up the kitchen in the morning, he occasionally finds the taps running in the men's bathroom. It can't be the water pressure, as it doesn't happen at any other time, just very early in the mornings.

People have heard mutterings and conversations in empty rooms. The tearoom is made up of three rooms, all connected by low doorways, but none with actual doors. Sometimes, if you stand in the bar, you can hear a low murmur of voices in the right-hand room, but if you look around the corner, the voices stop and there's no one there.

I can just hear the sceptics pointing out that it'll probably be just the central heating cooling down, or the pipes heating up, or the brewery equipment chuntering to itself. But the noise is distinct from the sound of the heating or the boiler. It's the definite sound of a conversation, but just low enough that you can't make out any actual words.

Outside the brewery is a long passageway that connects to the barns at the back of the farm. Originally the passage didn't exist, as it was part of an enclosed farmyard. At one end, there's a smaller, low-ceilinged room that the brewery staff use as a 'bonded warehouse', to store all the beer and casks before they are delivered. Some staff have seen people-shaped shadows in this room, slipping past the door and the windows or standing in the corner. One lady suddenly smelt a very strong, very unpleasant smell – 'like a thick, choking sewage smell', which disappeared as soon as it appeared.

The brewery buildings were built in 1840 – although there has been a farm on this site for a very long time. The buildings we converted were originally granaries, stables, cart sheds and hay barns. Perhaps the previous occupants don't approve of their new function as a tearoom or restaurant?

This sort of stuff doesn't worry me at all. If I'm wrapped up and not too cold I find working outside in the dark calming, especially on bright moonlit nights when you can see the animals calmly snoozing or chewing their cud. I've never actually seen anything, but I've certainly heard the rumble of low conversation and the sound of something running from one side of the building to another. Very strange.

Steve has also seen some weird stuff. Not on this farm, but where we used to live. We rented an eighteenth-century cottage for a few years before Steve inherited High House. Steve was sitting at the kitchen table in this cottage when he suddenly stood up and pushed his chair backwards. He looked like he'd had a fright and went a sort of strange grey colour.

'What's the matter?' I asked.

'I've just seen a bloke walk through the wall,' he replied, pointing at the flat kitchen wall behind the cooker. 'Bloody hell. I could see what he was wearing. He just came in and walked through there,' he said, gesturing towards the cupboard.

Apparently it was an older man wearing a lighter top 'with a sort of waistcoat over it' and dark trousers. He had white hair.

I had to give him whisky after that to get a bit of colour back into his face. Steve also saw the man standing next to the window in our bedroom, as if he was looking out into the garden. 'He was smiling, and I could see his upper body clearly, with a dark waistcoat and a collarless shirt. He looked happy.'

Back to the brewery.

Heather was approached by a local 'ghost-hunting' group that wanted to have a look around the buildings for Halloween. I asked if I could join in, as I'm a nosy old bag and wanted to see if they'd find anything out.

When the group turned up I was cheered to see that it was led by a man called Brian who was wearing an amazing orange wig. He had normal brown hair round the side, but sticking out the top of his head like a frill on a pie case was a bright ginger toupee. He was also wearing a necklace with an Egyptian 'ankh' symbol on it, which he wore because apparently in a past life he'd been an Egyptian priest.

Fair enough. If you want to go around saying those kind of things then best of luck to you. Especially in rural Northumberland. I thought the evening was off to a brilliant start, and tagged along at the back of the group eagerly.

We all trooped off into the brewery kitchens, with Brian almost immediately saying that he'd picked up the name 'George'. Now almost everyone was called George in the nineteenth century round here, so I didn't think too much of that one. Then Brian went quiet and said he saw a small ragged boy hiding in the side of the hay shed. The boy couldn't understand what was happening at the brewery, and why the old farmyard was full of bridal parties and visitors in big hats. I started to feel sorry for the boy, until Brian mentioned that he lived in Roman times and wore a tunic and sandals. I started to think that maybe Brian wasn't telling the truth. I can't believe that *anyone* wore a skirt and sandals round here in the fifth century or ever. They would have been bloody freezing.

Not much else for a while, apart from one of our party going into hysterics as she said she could feel a dog licking her hand. We had to guide her into the tearoom and sit her down with a hot tea until she calmed down.

'It was horrible! It was definitely a tongue! And it licked me!' I still don't know what that was. Imagination? A huge rat? Sumo the farm cat?

At the end of the evening everyone stood expectantly in the barn while Brian intoned a summoning to any spirits hanging around the farm.

'If you can hear me, come closer to us! Do not be scared. Give us a sign that you are with us! We are here to help you move closer towards the light!'

We waited for a response, holding our breath, ears straining for any reply.

Suddenly, we could hear faint crunching and dragging sounds coming towards us from outside. They got louder and turned into definite footsteps, limping over the ground, as if the invisible walker had an injured leg and was dragging a foot across cobblestones.

I grabbed the lady next to me, and she grabbed onto Brian. The footsteps got closer and closer, and Brian whispered hoarsely, 'Show yourself spirit! Come closer to our circle!'

After a few seconds a bemused face poked around the side of the stone wall. It was my neighbour, wondering what on earth we were doing in the shed in the middle of the night.

We all coughed and, avoiding each other's gaze, pretended to examine the floor and rafters in the torchlight.

Almost immediately after that, Brian and his wig packed up and left. I hope he comes back soon.

Wednesday, 1st November

It's a friend's Halloween party tonight. Everyone is dressing up as sexy cats and vampires and things. I'm going as a small, fat, grumpy farmer's wife.

Thursday, 2nd November

Today Candy the fat pony has disgraced herself by biting a small child and hiding behind the gorse bushes at the bottom of the field.

I put her bridle on and a little friend of the kids adorned it with a pink unicorn horn. Candy looked at me with utter disgust and is now sulking behind the gorse bushes at the bottom of the field, flirting with the expensive hunters over the wall.

I really hope that when I'm as ancient as her I have her unshakeable confidence in my saggy and decrepit charms.

Friday, 3rd November

I have new waterproofs! They fit me properly for once, so I no longer look like a hand-me down. Buying a new coat got to my head and I went a bit mad in my favourite shop (Robson & Cowan's agricultural merchants in Scots Gap). I also bought some new wellies, as my old pair started to leak, and I'm now the proud owner of a pair of Aigle welly boots – which are wide fitting to wrap round my tremendous muscular calves.

I showed off my new wellies by marching up and down in front of the brewery staff, and it started a long discussion about benefits of various welly brands. The consensus was not to buy the posher brands, as they fall apart after six months, but to stick to the straightforward 'Dunlop' brand, which lasts for ages and seems impervious to cracks or holes. Steve refuses to buy lined boots, as after wearing them all day in hot weather they get very stinky, especially if he's doing a sweaty job like shearing sheep or mucking out lambing pens. Although Dunlop are great, I much prefer a more flattering posh boot as there's nothing like a welly that flaps around your calf to make you look like you've just escaped from the local madhouse.

Saturday, 4th November

A brand-new pond has appeared in the wettest part of our recently planted barley field.

Steve has just noticed it has seagulls floating on it, as if it was a proper, permanent lake. He's now marching around the house shouting about the weather.

Sunday, 5th November

Today it's pouring down with heavy, continuous rain. The ground is so wet that when you drive the quad bike the grass ripples in front of the wheels, as if you were riding over a bog. All the hedges and trees are dripping wet, their bark and leaves dark and glossy with water. The ewes look sodden, their fleeces grey and heavy with rain. Every few minutes they shake themselves like wet dogs and shower the grass with water droplets.

Thrusty and Randy the tups have gone to meet their harem of ewes.

The kids have eyes like saucers, and I reckon there is no better way to teach the facts of life to two Northumbrian children than watching a randy tup work his way through a flock. Lucy was completely matter of fact: 'Dad, look at Thrusty – he's working really well!'

Thrusty was balanced on top of a lugubrious ewe, with his ears flapping backwards and forwards energetically.

'Why are they having piggybacks?' asked Ben.

Thrusty and Randy have been given around forty-five sheep each to service, and we all stand out in the bucketing rain for a few minutes, furtively watching them 'work' through the flock. Some of the ewes are very interested in the tups and follow them around, sniffing their fleece and bottoms. Others are completely uninterested and refuse to even look at the boys, grumpily running off when Thrusty or Randy come up to say hello. We leave them to it and go back inside the house, shedding a layer of waterproof jackets and hats and leaving them to steam dry by the radiator.

Monday, 6th November

Today Thrusty is looking ill. The terrible weather, plus the exertion of chasing around a flock of grumpy ladies, has given him a cough and a wheezy chest. Steve has diagnosed pneumonia, so the kids and I herd all the sheep into the pens while Steve catches a feeble-looking Thrusty and gives him the most enormous injection of antibiotics I've ever seen. It goes straight into the muscle on his bum, and he starts to perk up almost immediately.

After all their hard work tupping over the next couple of weeks, we should hopefully have lambs in the first week of April ... fingers crossed.

Wednesday, 8th November

This morning Steve trudged off up to his field of oilseed rape, to put up the 'Terror Hawk' bird scarer: a hawk-shaped kite on the end of a very long piece of string.

Steve spent a good twenty minutes untangling the string – it reminds me of Christmas tree lights. No matter how carefully you put Terror Hawk away at the end of spring, he tangles himself up into a tight ball for the next season.

The point of Terror Hawk is to stop the pigeons from eating the crops. What they do is peck out the middle of the baby oilseed rape plants, and it stops the little plants growing into bigger plants. Every year we declare war on the pigeon army that descends each morning and evening to decimate our fields.

We have a noisy bird scarer (or 'bangerbanger' as it's called by the kids) that we use on the bigger wheat fields. It's a gas-powered cannon that goes off with an alarmingly loud retort (like something out of World War II) every forty minutes during daylight hours. We carefully position it as far away as possible from the road, and point it away from houses, but it's so loud that any horse riders in the vicinity tend to spin round and go off down the road like the clappers. We're all used to it now, but it's a bit of a trouser filler if you're standing next to it.

Friday, 10th November

I'm sitting on my garden wall in the bright autumn sunshine, with Mavis our border collie.

She's beautiful, small and slim with a dainty pointed face and a sleek, long black coat that curls down over her white belly and feathered legs. We bought her from a lovely lady on Twitter, who named her after her Welsh Granny. Mavis has impeccable pedigree and is absolutely worth her weight in gold on our farm.

She's very gentle, with big brown soulful eyes and a long fringed tail that curls up on itself over her back.

Today she's filthy, as she's been helping Steve check the flock and it's deep in mud in some places on the farm. She flops by my side panting happily then potters off, tail wagging, to the field fence to stare at the ewes. They stamp their feet and stare back.

Mavis could never be a family pet. She's too focused and obsessed with working sheep. She runs about 4–5 miles every day alongside the quad bike. She doesn't like riding on the back, preferring her daily exercise no matter how wet and muddy the ground – unlike our previous collie dog, Nel.*

Mavis likes people but isn't interested in pats or strokes. She'll come and say hello but will then slope off to find some sheep to stare at. That's what she lives for. I used to try and

---

* Our previous collie dog, Nel, used to refuse to get *off* the bike when checking the sheep, and used to sit on the leather seat to survey her flock from afar, especially when she was getting on in age. She would also follow any walkers who used our footpath and sit and stare at them until they gave in and shared their packed lunches. She was much loved and eventually died of old age, and is buried under the clematis bush next door.

make her stay in the house with me while I worked, but she hated it, and would sit on the mat in a sulk, occasionally hurling herself against the closed door while making a constant whine/grumble.

Mavis lives in her own cosy kennel and run, which is inside the main passage of the sheep shed. She's warm and dry there in her basket and can supervise any poorly sheep that might be recuperating inside. We have given her dog toys but I'm not convinced she plays with them, as they usually sit ignored at the far end of the run.

Her breeder said that right from an early age she had an amazing 'eye' and instinct for the sheep, and while other pups would roll and tumble with each other she would prefer to sit and watch the flock.

She's a beauty. And priceless when we move our ewes. In the interregnum between Nel passing away and saving up to buy Mavis, the kids and I had to belt around the field to gather the ewes, and it was exhausting. Mavis nips and darts around the sheep, and is fast enough to turn back a runaway animal or cut off a stroppy ewe trying to make her escape. We couldn't manage the flock without her.

I give her a whistle and she runs back towards me, but then spots Steve in the distance and shoots off to see him instead. She's really Steve's dog. They've built a strong bond, and she much prefers to work with him, especially as he's more likely to be doing something interesting with the sheep flock. I can see her in the distance lying a few feet from where he's tinkering with the tractor, her eyes trained on the ewes in the paddock.

Sunday, 12th November

It's a clear, bright day, and Lucy and Ben have just disappeared off to their 'favouritest' place on the whole farm: a den next to a very old ash tree at the bottom of the front field. The tree has grown around the field wall, and is so thick around its widest part that it must be at least 150 years old.

Next to the tree is a deep translucent pool that has frog-spawn in the spring and clouds of tiny minnows in the summer.

Lucy has built a pretend sleeping platform in the ash tree, and Ben has his den underneath the holly bush beside the pool. They spend hours down there, carefully collecting berries and leaves and sweeping their dens clean with bunches of twigs.

Today I've given them a packed lunch and a drink and Lucy has taken her mobile phone. I'll see them in a couple of hours, probably with wet feet and muddy trousers.

Tuesday, 14th November

Candy is in disgrace again. She spent all night carefully un-tying the gates around her pen, undoing the shed door and letting herself into our feed store. She then spent hours chewing a very small hole in one of the feed sacks until she could suck sheep nuts out one by one, like a small but very powerful hoover.

As soon as Candy heard us coming this morning she whipped back into her pen and stood behind her hay net, pretending she'd been innocently standing there all night.

There were cross words: 'Bloody hell! These bags cost a tenner apiece you know! I'm not wasting them on that blasted field ornament!'

The field ornament pretended to eat a bit of her hay net, but her bulgy tummy was a giveaway. We've got the vet coming on Friday to do her teeth, but also to check that the extra feed hadn't done anything terrible to Candy's insides.

Steve is less convinced. 'Perhaps it's for the best if she did conk out! She doesn't bring any money in. If she did something, then fair enough, but she just eats her head off and farts a lot.'

Thursday, 16th November

Robert is a retired farmworker who lives with his mother in a local village. He's never married and has made a living doing odd jobs for local farmers in the district. He's now in his 60s, and every day, whatever the weather, he walks the road between Stamfordham and Corbridge that goes past our farm. When you see Robert striding around the corner, thumb stick in hand, flat cap jammed down over his brow, the best thing to do is to leap back into the hedge or duck behind a wall. He's a lovely chap, but the biggest gossip this side of Hexham.

All the local farmers know him and have various 'Robert avoidance' techniques. I've seen Steve duck behind the wheelie bins to try and escape.

Today, I got completely stuck.

Robert had caught me as I was walking back to our house and he settled down on the farm drive for a nice long gossip. He was very animated, as he'd spotted our new tups in the front paddock and wanted to know exactly how much we'd paid for them, who they were from, and how they were doing. Robert is especially interested in the price of things, and who has paid under or over the odds for beasts or machinery.

As we were chatting, our next-door neighbour drove past and saw us both. He gave a sarcastic thumbs up. Bastard.

Robert and I eventually got on to subsidy payments, and how much he thought all the neighbours had received that year. He was angling to find out how much our Single Farm Payment had been, but I managed to head him off. At one point I was going to offer to fetch our farm accounts so he could see for himself how we were doing.

Eventually I was rescued by Steve, who had seen me outside the window, took pity and gave me a call to pretend that he needed my help in the sheep shed. Robert cheerfully strode off, probably straight to the next farm to tell them all our news.

According to local lore, Robert and his mother don't have a television or the wireless, but prefer to spend the long, dark winter evenings cutting up the *Hexham Courant* to make sheets of toilet paper for their outside netty.

Saturday, 18th November

Sumo the farm cat and I are in a standoff. He refuses to eat anything but very expensive gourmet cat food. I'm telling him he's a farm cat and he'll eat Whiskas. He's just done a dirty protest outside his litter box. I'm not giving in.

... I gave in, after hours of endless yowling for food. Sumo is now tucking into a pouch of cat food described as 'the best quality responsibly sourced cod, marinated in an aromatic sauce with tender vegetables for optimum feline health'.

This is the cat who regularly kills and eats full-grown rabbits on our front doorstep.

WINTER

Traditionally, winter is the quietest time of year, when dark nights and cold weather stop us working long days outside. However, this is also the most vulnerable time for our ewes, who are pregnant and need careful checking and extra feed to ensure their unborn lambs are growing properly. Winter is a chance to sell last year's lambs at the Mart and (hopefully) make a bit of profit and prepare for lambing in the spring.

Sunday, 19th November

There are warnings of major winds and heavy snow in Northumberland. Which means I put Candy the pony out at 9 a.m., and at ten past nine she's standing pathetically by the gate asking to come back in.

The snow is heavy and has formed a thick white crust on Candy's back. I tell her that she's supposed to be designed for weather like this. She scrapes at a snow drift, exposing the grass below and starts to mournfully nibble on strands of grass.

She has a very thick coat, which is great insulation, proven by the fact that the snow currently heaped on her back isn't melting at all.

At 2 p.m. I look out the window and Candy is standing behind a bush, tail clamped tightly down, head and neck horizontal under an increasing load of snow. The snow is balanced evenly across her back and neck and an icicle hangs from the end of her forelock.

I can't stand it, and go out to call her in. She gives a welcoming whicker, shakes off the snow and plodges across the paddock for a sheep nut. I lead her inside her stable and leave her knee-deep in warm straw with a feeder full of fragrant hay. Before I leave I give her a lecture about her Shetland ancestors, who lived out all the year round, and ate anything they could find, including seaweed and patches of moss.

Monday, 20th November

The kids adore snow days on the farm. Today there's a heavy layer of white like the top of a wedding cake. After checking the sheep, we attach a plastic sledge by a long rope to the back of the quad bike, and then drag it after us around the field at top speed. Lucy and Ben adore it and scream and laugh at the top of their voices. It's probably a health and safety nightmare but if Steve drives the quad, I can perch backwards on the seat and make sure that the sledge doesn't slide under the wheels.

Ben keeps falling off when we slither around a corner, and it makes me laugh watching his pink cheeks and hat and gloves covered in ice and snow. Back home, everyone leaves their coats, scarves, boots, hats and gloves in a big sodden pile on the hall floor, and we all drink cups of tea and hot chocolate.

Tuesday, 21st November

When Steve is at work on a Monday and Tuesday,* Mavis and I have a nice long walk through the farm. We like to trundle down to our wood, and poke among the bracken and brambles to see if anything interesting has happened since our last visit.

Today I was trudging along the side of the wood, lost in thought, and I spotted something that looked like a crumpled-up blue sleeping bag snagged in the undergrowth.

I moved a little closer, and Mavis shot across my path, from left to right, dragging what looked suspiciously like a human thigh bone.

With breathless visions about murdered hikers, I managed to call Mavis to heel, and take a closer look at the bone. When I got closer I saw that it still had an articulated hoof at one end. Mavis proudly took me to the carcass of a roe deer, with the skull and a few scraps of hide still adhering to the bones. The blue 'sleeping bag' turned out to be an enormous deflated 'CONGRATULATIONS!' party balloon that had become stuck in a bramble thicket.

Mavis settled down happily to chew on her bit of dead deer, and I had a rest on a tree stump for a while to recover.

---

* Steve works part-time as a site manager to pay household bills and help bolster the farm income.

Friday, 24th November

We've started our Christmas shopping. Steve doesn't really get the excitement of the run-up to Christmas Day, but I absolutely love it. Our tree goes up on the 1st of December, and I run around the house throwing greenery over pictures and mantelpieces. Our Christmas tree usually looks like Santa has sneezed on it – there is no carefully thought-out colour theme here. All the tinsel and baubles are layered on, with the occasional cotton-wool-and-cardboard angel made by the kids at school.

Steve does enjoy shopping, especially for tools, machinery and technical gadgets. We very rarely get out to Newcastle, and when we do we tend to walk around in a daze staring hypnotised at the lights and sparkly decorations.

Today, we've been shopping at Fenwick's store in Newcastle city centre. Steve keeps shouting 'look at the shiny things!' and dragging me to stare at displays of headphones and other techie gadgets. It's bedlam; I've never seen so many shoppers. Random people keep treading on me. I discover from one of the shop assistants that we've chosen to come shopping on Black Friday, the busiest shopping day in the year. That explains it all. Exhausting. We manage a couple of hours then drive home, breathing a big sigh of relief when we hit the A69.

Saturday, 25th November

Slight problem today. The heating has gone on the blink. There's no hot water and the house is icy cold. Steve crashes around upstairs shouting about stopcocks. We quickly light the wood stove and settle the kids in front to get into their school uniforms and eat their breakfasts.

I can't imagine how difficult it must have been for everyone to keep warm before central heating. When I was younger my parents didn't have radiators, and we lived in a tall, thin, four-storey Victorian house with big rooms and high ceilings.

I well remember the mad dash from warm sitting room to my icebox bedroom, clutching a hot-water bottle before cuddling down below the icy sheets, curling up round the heat and trying to keep warm.

Originally our farm cottage used to be two rooms – one up and one down. It must have been warm at least, looking at the enormous fireplaces with the heavy stone lintels. Where would they have stored the wood? How did they keep the fire on all day? Our outside shed was originally the netty, so it must have been a freezing dash to go to the loo. How did they keep clean? Maybe they didn't. I'm certainly not washing anything more than my hands when there's no hot water.

Steve is now staring into the fuel tank and has discovered that although we've ordered the oil, the delivery man hasn't been, so the tank is empty. We wait on tenterhooks all day, looking out for the tanker, who has promised to get to us before nightfall. Of course, the temperature barely rises above freezing so by 4 p.m. I'm wearing a vest, two pairs of Steve's shirts and a huge, shapeless jumper.

Monday, 27th November

The heating oil has been delivered and the radiators are blasting away. I'm having a bad day, so I've wrapped myself up in my old flannel blanket and am lying under my duvet. I haven't brushed my teeth or hair and I'm not answering the phone.

A lot of people are talking about mental health and farming on social media and the radio. I've had struggles with anxiety since I was very young, but it's taken me until my fourth decade to speak honestly about it. It's hard to admit to problems when you've tried to mask them for a long time. But constantly pretending that everything is OK can be very, very tiring. I don't want to sound like I'm indulging myself, but it's been bad lately, and the usual ways of squashing it down have stopped working as well as they used to.

I worry about everything and anything. Sometimes I worry

that I haven't got much to worry about. And sometimes I worry about how much I'm worrying and worry that I'll end up with some awful disease and worry myself into an early grave. I do wonder if by the time I'm 60, I'll be wearing a tinfoil hat and gibbering about aliens at the postman through the letterbox.

The things that do work for me are: exercise (although I really, really hate it and avoid it as much as humanly possible); lying on the carpet wrapped in a blanket; sitting in a patch of sunlight; talking to the fat pony; having a break from social media; seeing my mum and friends; reading many, many books; and not rushing around too much. Mindfulness also helps, although I force myself to do it. I have a nifty little app on my phone that bings to remind me to do the practice each day. It stops me being so fidgety and cross and doing weird stuff such as cleaning under the bath with a toothbrush.

Things that don't work are: having long, imaginary conversations with people who once said something unkind to me ten years ago; eating a lot of sugar; ruminating over everything I have done wrong at work since 1999; thinking obsessively about being bullied at school; rehashing conversations in my head to check if I said something weird; and compulsively refreshing Twitter or reading the Daily Mail.

Once you've got excessive anxiety yourself, it's easy to identify other people with the same problem and the things they do to try to keep it in check. I'm very much in awe of those who have these problems and still achieve stuff: keeping a good job and a nice home and looking like proper grown-ups with brushed hair and teeth and everything. I can only imagine that sometimes they're just like me – lying under the duvet trying to speak sternly to themselves.

I've always been told that only the best type of people get anxiety – the most intelligent, high-achieving, Type A personalities. Or maybe that's what my GP tells me to make me feel better. I prefer to be friends with people who are thin-skinned and worry a bit too much. People who appear to breeze through life make me very jealous. But maybe they're also fighting their own hidden demons that I know nothing about.

Anyway, I've managed to give myself a kick up the arse and got up to write this diary entry. It feels nice to finally talk about it. I like my life a lot, but sometimes it just gets a bit too much. But I count my blessings that I'm still fairly sane, have lots of support and help and can see the funny side (sometimes).

Tuesday, 28th November

I am still tired today but feeling a bit better after doing bugger all yesterday.

I'm forced to get up early, as at first light we had a phone call from Katie, who lives next door.

'I've just seen a small, grey pony trotting up the road. I think it might be yours?'

Bugger. Of course it's ours ...

We stagger outside (in −7°C/19.4°F) and eventually find the fat pony, standing in our next-door neighbour's farmyard, staring thoughtfully at his cow shed.

She is very pleased with herself. Mavis the collie and I herded her home, and she trotted along the road, throwing in the odd farty buck.

As she has pretty bad arthritis, she's finding it hard to get around and bend her front legs. She's been prescribed horse paracetamol (phenylbutazone) to help with the pain. It's like pony speed, and it's given her a new lease of life, and she's now happily bobbling round the farm, undoing bolts and getting in the way.

Last night she'd managed to undo *two* locked stable doors and, judging by the small pony hoof tracks, had been to visit the sheep, Joey the hunter, our hay shed and then decided to branch out and visit next-door's cows.

She's probably eaten 3 million calories in high-quality cattle silage.

Tonight I shall be shutting her in with the locks tied up with baler twine. No doubt she's already planning her next escapade ...

Wednesday, 29th November

The cold weather has made me break out my comfort food recipes. Today I made liver and bacon for lunch … as it's good for you. Now the whole family is hiding upstairs, refusing to come down and pretending they're not hungry.

Steve loathes stews, casseroles and baked potatoes. Ben won't eat anything 'with bits in it', or 'sensibles' (he means vegetables), and Lucy hates pasta, chicken and fatty beef. I don't like fish pie, beans or avocados.

Trying a new recipe is a bit of a minefield. Eventually I bribe everyone into trying dinner by promising ice cream for pud. They all pick over their plates and finally compromise by eating the bacon, mashed potatoes and broccoli and scraping the liver into the bin.

Friday, 1st December

I ring my mum every day and see her about three times a week. I love her to bits. She's a smidgeon over five foot, hates cold and wet weather, laughs uproariously at the ridiculous things in life, and has spent her entire life looking after Dad, my brother and me.

Today she's come for a cup of tea and a natter, and is now on at me for not stocking my fridge with the right food.

'Look at that ham. You always get the cheapest ham. It's awful ham, all thin and anaemic-looking.'

'But Mum, we like that ham. I have it every day in sandwiches,' I retaliate.

'Why don't you cook a proper ham joint? Then you can carve a bit off each day. It's much nicer for the children.'

A few years ago, when the kids were small, she was convinced that I should 'boil a chicken' every week, which would give us all nutritious broth and a way to feed the family every day.

'But I don't want to boil a chicken, Mum,' I used to say plaintively, while refusing to open the door of our fridge so she

69

couldn't see what was inside as she made disparaging comments about my shopping ability.

'Oh, you always get the most boring shop,' she often says, while inspecting my weekly delivery, as if she goes shopping at Harrods delicatessen each week. But then she'll make me pies and cakes and bring them up with a 'I thought this would be useful' comment. They are always needed and much appreciated, especially when I don't have time to bake or cook proper dinners.

I'm not exactly skilled at cooking. My mashed potatoes are never fluffy and my Yorkshire puddings are flat as a pancake. Steve is a much better cook than I am, except he lays everything out at the start of each cooking session, all split into little bowls, with his tools lined up on the bench. I find this infuriating.

I tend to just go for it, and hoy in the ingredients that I think will work, as I'm too bored to check the recipe. Sometimes it works and sometimes it doesn't. If I won the lottery, I've decided that I would need to invest in a chef so I'd never have to cook chicken nuggets or baked beans again.

Sunday, 3rd December

I'm still eating too much, not losing any weight, and today I've decided to give up going to the slimming club in the village. Dad still goes and has lost about a stone. I'm all depressed about it. I can imagine him headlining their next newsletter: 'LOCAL PENSIONER LOSES HALF HIS BODY WEIGHT!'

And in smaller letters, 'HOWEVER, DAUGHTER STILL LOOKS LIKE A BLIMP!'

He loves the meetings, and now has a small coterie of ladies that laugh at his jokes (which seem to be the same ones each week) and cheer him on when he's managed not to eat his usual cream scone on his weekly bike ride.

Monday, 4th December

I'm checking the sheep in the back field. This flock are made up of ewes that are two years and older. Some of them are much older, and some of them are bordering on the geriatric. My favourite sheep are in here: Mabel, the tea-coloured ewe; Spotty Nose, an old distrustful matriarch; and Fatty, the enormous oval-shaped sheep that produces one spindly lamb a year.

While I'm walking down the side of the field I spot Mabel standing on a hillock stuffing her face with grass.

'Mayyyyyyyyyyyybeeelllllllllllll!' I holler at her.

She picks up her head and stares across at me, squinting her eyes against the winter sun to see who it is. She recognises my face,* then starts lumbering towards me over the riggs. Some of the furrows are so deep she disappears from view at the bottom before popping up again on the other side.

When Mabel reaches me she's panting like a pair of bellows, and starts snuffling in my pockets for sheep nuts. I always have a supply, and I'm so chuffed that she ran across to me that I give her three huge handfuls.

She gulps them down then reaches up her head and delicately sniffs my face. All sheep seem to do this. It seems to be their greeting/recognition system. I huff back at her, and then spot Fatty galloping across for her portion of nuts.

Fatty is round like a beach ball and isn't as athletic as Mabel. Her galloping is more of a wobbly canter, with her belly swinging from side to side with each stride.

She plunges her nose into my pocket and Mabel steps to one side as if she's embarrassed by her greedy flockmate.

Spotty Nose won't come over. She's very suspicious of all humans, and remains aloof even after some fairly intense

---

* The latest research says that sheep can recognise up to fifty different human faces. Our flock can definitely recognise our family members and are much more wary of strangers, refusing to come to say hello until they're tempted with sheep feed or neck scratches.

'sheep-taming' sessions involving sheep nuts and the odd apple core.

After some time petting Fatty and exclaiming at the width of her arse, I wander back home. Afternoons like this are glorious. It's frosty but dry and still, and I can hear the flocks of jackdaws chattering in the trees in the hedgerows. It gets dark now just after 3 p.m., but this evening there's a stunning sunset, all vivid pinks and glowing oranges. It usually means it'll be a bright day tomorrow, with at least some watery winter sunshine.

Thursday, 7th December

I'm doing the late-night check on the animals but I've forgotten my torch. Luckily it's a full moon, and the light is so bright that I can make my way around without ricocheting off any projecting stone walls.

The moon is low on the horizon and looks bigger than usual. It's so clear that the field fences are throwing 'moon shadows' over the frosty grass. The stars are bright, and I can pick out Orion's Belt and the Big Dipper. There's hardly any light interference here, and although I can see the pinkish hum of light from Newcastle on the horizon, the skies above our farm are completely dark and perfect for star-watching.

Looking at the sky makes me feel very small, and I start to think about the generations of farmers and farmworkers that have done the same as me every evening and walked the same tracks and fields. Some of our field footpaths have high banks, where the track has been worn down by generations of farmers and labourers checking their stock and crops.

I've read somewhere that there are signs of Hadrian's Wall being built directly over freshly ploughed fields.* I'm convinced that the footpath running from our farm to the wall

---

* That would piss me right off. There you are, having laboriously ploughed out your small wheat field, when some Italian comes and builds a bloody great wall over the top of it. So much for your harvest.

is much older than we think; the farm might have even been there when the wall was built.

I love learning about the 'small' history of a place. What happened to the peasant farmers and families, trying to raise children and grow enough crops to tide them over through the colder months?

There's an abandoned medieval village around the corner from us, with a big village green and lots of little rigg and furrow strips in-between each bumpy house plot. I've read the Historic England report, but it's a dry document, all about church lands and local big-wigs. It seems the villagers didn't abandon the village due to the Black Death but were 'moved off' by the local landlord and settled on other small farms. I want to know about the minutiae of their lives. Where did they go to church? Who brewed the village ale? Where did they wash their clothes? Where did they keep their sheep and horses? It must have been a huge undertaking to decide to move your hearth and home from a well-known farmstead to a new one. I wonder what they were like. I wonder what their animals were like. Probably miniature sheep and the odd scrubby-looking horse.

I try to imagine these long-ago people. Sometimes, while walking around the farm on a frosty moonlit night, you feel as if they are walking behind you, or to one side, just out of eyeline and earshot. I wish I had the benefit of their generations of experience and knowledge of animal husbandry. Shivering, I finish my late-night check and go back into the house, shutting the cold night out behind me.

Sunday, 10th December

My brother has come to visit. He's older than me and works as a GP in Yorkshire. He's thin as a rake, is desperately worried about the NHS and is the funniest person I know. He's very good at his job and has been promoted to training student GPs or whatever you call apprentice medical students. His latest complaint is the fact that I never go down to see him and his family.

'But I have so many animals to look after! I can hardly schlep down to your house with a car full of chickens, sheep and the dog, just to come and see you,' I say.

'Come down for just a weekend and we'll go to lots of second-hand bookshops and get a curry,' he replies.

That's very tempting. My brother is married and has two sons. Alex, the eldest, is 14, and I'm trying to get him to come up for his Easter holidays next year, so he can help me with the lambing. I'm not sure if he's convinced. I keep texting him desperately jolly messages, such as 'You will enjoy it alex. It will build character', and eventually 'I will pay you money. And feed you lots of nice things'. So far, I've not heard a squeak back.

My brother works in a very busy practice, and the number of patients he sees each day is staggering. He starts telling me about his World Swivel Chair record, which involves seeing how many times he can spin round on his chair between each patient. So far he's managed fifteen rotations ...

Later in the day I find him in our wood, sitting on a tree stump enjoying the peace and quiet and generally communing with nature.

'There's a blue sleeping bag over there with a dead person in it,' he tells me warily.

You'd think a doctor would know the difference between a person's thigh bone and a deer leg.

Monday, 11th December

The holly trees and bushes are bursting with berries.

'Aye, it'll be a hard winter,' prophesises Robert on his daily walk past our farm, 'if there's loads of berries it means that we'll have deep snow.'

The frost is already cracking the puddles and making the farmyard into an ice rink.

All the outside troughs are frozen, and the inside troughs are beginning to freeze too. Making sure the animals have water is now our main priority.

Outside, the cold makes the plastic troughs fragile, and Steve warns me to be careful when cracking the ice. Except it's so thick that I have to hammer a hole in the ice using a pointy stone; after bashing it half a dozen times the ice is still completely solid. I end up climbing onto one trough and jumping up and down to make a hole big enough for the sheep to drink.

One morning it was –12°C (10.4°F), and my just-washed hair started to freeze under my hat.

This morning, the fat pony is standing sadly next to her completely frozen water bucket while the outside pipes are locked up with ice. I end up hauling tepid water from our kitchen across to the farm to fill buckets in the stables and sheep pens. It is hard work.

Later on I cut some sprigs of holly, but being careful to ask each tree's permission before I take off a branch, as per Steve's direction. Holly is supposed to ward off evil, so I put the red-berried twigs above our fireplace and windows and weave the others into a Christmassy wreath which goes on the door.

Wednesday, 13th December

This morning, when I feed the lambs inside the hemmels, Scabby the ewe hurtles up to the front of the pen and demands to be hand fed sheep nuts.

She's put on some weight, but I can still feel her backbone jutting out from her fleece, and her legs are as spindly and thin as twigs.

When she eats from my hand she gobbles as fast as she can, but she still spills many of the sheep nuts. Her mouth is so deformed that she can't seem to pick up the nuts correctly, and relies on a scooping motion to suck them into her maw.

Ben is still very attached to her, and goes to sit in the hemmel with her, stroking her curly forehead and whispering sweet nothings into her fluffy ears. I'm not sure how she'll cope in spring if she can't chew the grass properly. I'm cross that we didn't spot her problems before we bought her, but it's impossible to check every sheep that you buy in a bigger flock.

After blogging about Scabby I receive an email from someone who lectures me about the role of a farmer as the 'upholder of animal welfare' and how money should be no object when ensuring every animal is looked after to the best of our ability.

He's never been near a farm in his life. I want to reply and tell him that we still need to make enough money to keep the farm as a viable business. That of course animal welfare is at the top of our agenda but, yet again, we don't have unlimited pots of money to spend on the health of one animal.

I know plenty of farmers who would quietly put down Scabby without a moment's thought. Steve and I can't bring ourselves to do this, so Scabby continues to hoover in expensive sheep nuts and fail to get any fatter. I'm in a quandary. She can't be put in lamb, and she can't be sold at the Mart. The options are: a) keeping her as a pet and feeding her concentrates all year round or b) asking the knacker man to put her down.

I'm not making that decision today though. Scabby follows me around while I do my other jobs, and as I leave she's sitting in deep straw, contentedly chewing her cud.

Friday, 15th December

I've bought some very sharp foot-trimming shears this week, just in time for our next feet session on the fat lambs.

Since we've brought them inside, quite a few of them are limping, and it looks like they have foot scald or foot rot.

This is very common, but harder to treat than you think. Sheep need regular feet trimming as the deep cleft in their divided hooves provide a perfect place for bacteria to grow. Feet scald develops when bacteria is trapped and starts a painful infection.

I'll have to cut away all the excess horn on the foot and expose the infection to the air before spraying it with antibiotic foot spray and giving each affected lamb 10 ml of penicillin.

Foot rot smells disgusting, like a mixture of rotting cabbage, dead bodies and sewage. It lingers on your hands for days, even though I scrub them repeatedly afterwards.

I catch each limping lamb and tip them over onto their hindquarters so they can't struggle and I can reach their feet. It's quite an art, as they kick and flail when I balance them on their haunches. Steve catches and tips his sheep in one fluid movement, while it often takes me two or three times to properly position each animal. I tip one little lamb and she immediately slumps to one side, as if asleep. I wish all of them did this, as it would make the trimming so much easier.

Some of the worst-affected animals absolutely stink and their feet are hot to the touch. Clipping their hooves often means I pierce a pus pocket, and stinking yellow matter bursts onto the leg and drips down into the straw, while I almost keel over with the smell.

Official guidance says that badly affected animals should be culled, and feet trimming should only be done once or twice a year. In an ideal world that sounds like a great idea, but for a smaller farm like ours it's not really practical. If we culled each animal that had foot rot we'd be left with about twelve sheep!

The worst-affected lambs are the ones we bought in the Mart in September. They haven't yet managed to build up the correct antibodies to the bacteria that live on our farm, so they suffer more with foot rot than our existing flock.

On one or two animals, the entire hoof peels off, accompanied by a fountain of pus and blood. Strangely it doesn't seem to hurt them, and after soaking their feet in antibiotic spray, injecting long-acting antibiotics in their bum and pushing them through the foot bath, the lambs hobble back into the shed fairly happily.

My new shears are doing their job brilliantly, although I have a few near misses when a sheep kicks and I almost take out my own eye. I have to bend right over to see the hoof, which means that one wrong move and I could stab myself in the face. It's a filthy, smelly job, and after going through the entire flock with Steve we strip off our trousers and coats *outside* the house, so the smell doesn't permeate through the entire building. I'm pleased that I've been useful, but I can't help noticing that Steve manages to trim two to three animals to every lamb that I treat. My back is hurting as well, from lifting the heavy sheep and tipping them over. Being little is an advantage though, as it means I have less of a distance to bend than a six-foot farmer.

Saturday, 16th December

Steve staggers into the house waving a bloody hand and making incoherent 'oof, oof!' noises. He's managed to crush his thumb and it's a mix of jagged nail and bloody pulp. He's a strange green colour, and as I start to look at his hand he suddenly goes very pale and slumps over onto the floor.

Shit.

He's too heavy for me to move so I drag him into a sitting position. Steve has a history of fainting when he sees blood. It's a bit embarrassing for a farmer. He's fine with someone else's blood, but if it's his, or even mine, he goes all woozy.

I examine his thumb. He's crushed the tip and the nail has come away. He comes round and, ignoring his shouts of 'I don't want to go to the hospital!', I haul him into my car and drag him to the Emergency Care department at Hexham Hospital.

They greet him like old friends. In his life, Steve has split open his scalp twice, almost cut off his fingers a few times, gouged a hole in his leg and run over and broken his foot.

Farming is a dangerous business. We all work on our own, and deal with heavy machinery or large animals, and tiredness or inattention can cost someone their life.

We all know the horror stories: getting stuck in potato harvesters, or sucked into combines, or mangled by headers, or trapped under quads, or run over by tractors or the plough, or falling into slurry pits, or knocked down by a reversing loader, or crushed by rampaging cows, or falling off the top of straw stacks, or crashing through unstable shed roofs.

Steve has scraped together some cash to buy himself an insurance package. It's not a very cheery subject, but if he's badly injured, or god forbid, even killed, then there's no work payout or sickness package to fall back on. So each month he pays (quite a large) amount to ensure that if he did end up with one leg or no arm, he could still keep us all from the workhouse.

The full story emerges. He was hitching up the tractor to the plough when the bar got stuck. He got out of the tractor cab to have a look, and in a moment of stupidity tried to move it himself. Of course, the bar slipped and landed squarely on his thumb. Cue blood and bits of nail everywhere.

The nurses listen sympathetically, then stitch him up, update his tetanus injection, wrap his hand in a bandage and send us home.

We know a lot of farmers who have married nurses. Maybe they spend so much time in the local A&E departments that it gives them time to start chatting up the staff.

Back at home Steve promptly strips off the bandage and goes out to rehitch the plough, while I shout instructions at him to sit down and rest. He tells me he hasn't got time to be ill. He's continued working when he had pleurisy and stomach ulcers. God knows what it would take for him to stay at home.

Monday, 18th December

It's the week before Christmas. Steve has bought a beautiful six-foot Christmas tree from the sawmill in Hexham and sets it up in pride of place in our lounge. It looks fabulous, and it's lovely to have a proper tree.

But when I look closely, I realise that he's used an old sheep lick bucket filled with sand as the tree base. It's still got traces of old lick crusted around the side. I could make him wrap it in tinsel but instead decide to live with it. No one mentions the fact that our Christmas presents under the tree are around a bucket proclaiming 'Crystalyx Hi-Mag Molasses are No.1 for Sheep'.

Wednesday, 20th December

In the summer I had enthusiastically bought the family tickets to a special 'Christmas Evening' at Beamish Open Air Museum. Today is the day – and the evening promises all things festive: Father Christmas; snowy tram rides; brass bands and carol singers; and a proper Victorian travelling fair. It sounds wonderful, and it will be such a treat for Lucy and Ben, as Christmas, sparkly fairy lights and feeling all cosy are at the top of their list of favourites.

When we arrive, the car's thermometer is showing an outside temperature of –5°C. However, being farmers, we've come fully prepared for the weather, and are dressed in a multitude of padded anoraks and fleece jumpers. In about five layers each, we waddle off to see Santa.

Santa has a fabulous Geordie accent, is seated in front of a roaring coal fire, and has a herd of real reindeer surrounding his grotto. He's everything I'd hoped for. After the kids have faithfully promised him that they will tidy their rooms and stop hitting each other in return for a Christmas delivery of a hamster and a kitten, we wander outside to sample a stall full of cinder toffee and roasting chestnuts. By this time it's seriously cold, and when I arrive in the Edwardian Village I'm scuttling between shops trying to keep warm.

After the village the kids beg to go to the Travelling Fair.

'Maybe we should go to the coffee shop instead?' I say weakly.

But Lucy and Ben have been looking forward all day to the fair, and we're going to go, no matter if County Durham is the same temperature as the Arctic Circle.

By the time we've walked there it feels about -20°C, and I zip up my anorak to the tip of my eyebrows and pull my ear-muffs right down over my ears to feel less frozen. I can't feel my feet and my hands are so cold the fingertips are tingling.

'I think I've got frostbite,' I tell Steve, who never seems to feel the cold as badly as me. 'Rubbish!' he says, 'it can't be that cold.'

There's an old-fashioned carousel, a gypsy caravan, some side games and, thank the Lord, a person in a 'Victorian urchin' costume selling cups of tea. I clutch mine and huff deeply at the warm steam, trying to defrost the tip of my nose. After paying a small fortune for the kids to have a go on all the sideshow games, everyone decides they want to ride the carousel as well.

The children and Steve choose their animals: a majestic swan, a dragon and a fairy carriage. The roundabout starts to move, so in a panic I heave myself onto the nearest wooden horse. They're quite far off the ground, and being short I find it tricky to get on. My horse is painted sickly pink, has a stupid grin and is missing one of its hooves. Carousel horses are not the easiest things to hang on to, and so I clutch madly at the central pole while the whole ride begins to revolve.

The kids love it, and so does Steve, who, in his fairy carriage, is protected against the sub-zero Arctic wind. On my pony I'm completely exposed to the cold air, and a few minutes into the ride my top lip has frozen onto my gums in a rictus grin and my grip on the pole is beginning to slip. There doesn't seem to be any stirrups. Surely Victorian carousel designers wouldn't have overlooked stirrups? I have a terrible vision of me falling frozen, with a solid thump, onto the painted boards beneath and being run over by the bright orange tiger behind me.

Gripping on with frozen legs and feet, I look down and see that my horse's name, Willie, is painted onto its neck in big pink letters. The music gets louder and seems to be a fune-real steam organ version of 'Pop Goes the Weasel'. I revolve on Willie for many long, unhappy minutes in the teeth of a freezing wind, lips pulled back in a terrifying, frigid grimace, trying to avoid the eyes of waving onlookers. Eventually the ride judders slowly to a stop and I half slide, half fall off Willie onto the icy ground.

'You looked a bit uncomfortable there,' comments Steve, as I cling onto his coat, trying to absorb some of his body warmth. I hunch over trying to keep my fingers from freezing, with one hand in an armpit and one in my pants. I'm convinced I'm becoming hypothermic and start to suggest everyone should walk back up the hill to the tram stop when the kids catch sight of the other ride.

'Let's go on the shuggy boats!' they shout in excitement.

'Oh god, not the shuggy boats,' I plead.

'Take your hands out of your pants,' says Steve. 'You look like you're fiddling with yourself.'

The kids, with many happy squeaks, climb into the shuggy boat, while I lean on the fence and fantasise about forcing my way into the gypsy caravan. I bet they have central heating in there. Or one of those electric heaters that blow out hot air. Maybe if I stand in that ticket booth I'd be able to defrost? Eventually, painfully, the shuggy boat creaks to a halt and we set off back up the hill, the children with pink cheeks aglow and Steve striding out purposefully, while I bring up the rear in a kind of frozen half crouch, my feet too numb to walk properly.

I ring my mum that night and tell her about my ordeal. 'Did you get diarrhoea?' asks my mum. It's a rule that members of my family are struck down by stomach problems when they get too cold. My mum calls it 'catching a chill', and after a winter walk, there's always someone languishing on the sofa, clutching a hot-water bottle to their guts.

I didn't get diarrhoea, but there were definite ominous rumbles on the way back in the tram. All the other families at the exit look happy and festive and excited with their visit. I just shuffle into the car and fart miserably, crouching over the hot-air blower with the seat heater on full blast all the way back home.

The kids spend the evening chattering on about Santa and

the rides. 'I loved it all,' says Lucy in excitement, while Ben sits on my knee and explains how it 'was the real Santa, Mummy. It was the absolutely real one cos I could tell by his beard.'

It was definitely worth it.

Sunday, 24th December (Christmas Eve)

As a young girl I remember the shivery excitement of Christmas Eve, laying out my stocking and waking up to a satsuma in the toe, and usually a brand-new Sindy horse and Silver Brumby book.

Lucy and Ben are no different. Lucy has been promised a kitten from a half-feral litter at next-door's farm, and Ben wants a hamster. On Christmas Eve we all go to the Christingle and crib service at Royal Church – an ancient, tiny building with walls covered in Saxon grave covers featuring the swords and shears and old crosses used to denote who is buried beneath. The interior is lit with candles and hung with holly and is utterly magical. Outside, there's a full moon and a clear sky, and the grass is already crackling with frost.

After the service we all troop home and cuddle up round the fire, stockings, mince pie and a glass of good-quality whisky all laid out for Santa. I've tried to spread the cost of present-buying this year to help with our parlous finances, finding one or two little gifts each month for the kids since August. Once the children are in bed we stuff their stockings and lay out the presents. It always makes me proud to see that they both have a decent pile. However hard the year has been, we always manage to sort out a proper Christmas.

Monday, 25th December (Christmas Day)

Ben wakes me up at 2.45 a.m.

'I think I've just seen Santa! I saw a light in the sky and

it was definitely the reindeers!' He's literally vibrating with excitement.

'Go back to bed darling, it's too early,' I croak, and he gallops back into his bed and clutches the duvet up to his nose.

He's back again at 6 a.m., and we stagger downstairs to watch the children rip into their pile of presents. An hour later all is open, and Steve is already knee-deep in batteries and screwdrivers, methodically constructing a laser shooting game.

Cinders the tabby kitten has arrived in a cat carrier and is in a special cat pen with a fluffy bed, litter tray, water, food and numerous crinkly cat toys. We've put her in the spare room and she's tucked deep inside her bed, curled up tight and ignoring all attempts at stroking her fur. She's a beauty, with a silvery grey coat with dark brown whorls. Her mum and dad are ferals from the farm up the road, but she seems remarkably calm and sensible. Her whiskers are almost as long as her body, and her eyes are a deep, fiery orange.

Jellybean the new hamster has arrived in a great second-hand cage I found listed in the local newspaper. His fat little body is buried deep in a pile of shavings. He's a lovely, dusty yellow with a pink nose, surprisingly large yellow teeth and a black fuzzy spot on his bum.

But farming doesn't stop on high days and holidays, and after admiring the new additions to the household, everyone rams on coats, hats and gloves, and goes out to feed the animals. Steve and I are holding our breath. It's passed into family lore that on Christmas Day – or Easter Day, or a bank holiday – there will be something the matter with one of our flock. They seem to take great delight in getting ill or dying on a day when everyone else in the country is lying in bed, eating Quality Street and watching telly.

Today is no different. There's a ewe with her head stuck through the fence at the bottom of the field. The kids, Steve and I bounce over the frosty grass on the quad and with frozen hands carefully work her head out from the wire strands. She doesn't seem to be any worse for her ordeal, and trots smartly away with a few bald patches around her neck.

Then, after feeding all the animals, bedding up with clean straw and smashing the ice in the water troughs, we can go back to the house and Christmas Day starts properly – with hot chocolate and a huge bacon sandwich.

Tuesday, 26th December (Boxing Day)

I thought about taking Candy to the traditional Boxing Day hunt but realised I would have to spend most of Christmas Day washing her to try and get the ingrained dirt out of her thick coat. And then I would have had to keep her clean overnight with some kind of home-made sheet, as I haven't got a rug to go around her stomach.

Instead, we pile down to Corbridge to watch the traditional Hunt Meet, to pet the hounds and stroke the horses. It's a cold, clear day and the horses' hooves ring out on the tarmac in front of the village Market Cross. Everyone is in tweeds and woolly hats, drinking hot mulled wine and eating sausage rolls.

Back home there's more feeding and bedding up. We use the tractor to distribute huge bales of hay to the sheep outside. The grass has literally no feed value at this time of year, so the ewes crowd round when we unwrap the bales. It has the most fantastic smell – part fresh grass, part floral/fruit smell, with a hint of bananas. Almost good enough to eat ourselves.

Sunday, 31st December (New Year's Eve)

I am not a fan of New Year's Eve. I can never keep my eyes open until midnight, and often fail to see the new year in altogether. Steve talks wistfully of country parties when local farmers would pile into cars (no drink-driving laws then) and steam off to others' houses to go 'first footing'. This involved a dark-haired man being pushed out the back door with an orange and a piece of coal. He would then knock, and amid much hilarity, the plastered partygoers would let him in and everyone would get back to the serious business of drinking. It was supposed to ensure the houseowner and their family enjoyed prosperity in the next year. Sometimes these parties went on until first light, so I can imagine that feeding animals and mucking out must have been hard work with a hangover.

Instead, we eat a takeaway, check the animals, put the kids to bed and go to bed ourselves, with Cinders the new kitten snoring happily on our feet. Before we go to sleep we raise our cups of tea and wish for a healthy year – happy kids, a good harvest, a good lambing and maybe a little profit in the bank.

Wednesday, 3rd January

Today, Steve and I decide that our flock of in-lamb ewes need to come inside to be foot-bathed for scald, as their feet are beginning to get sore again in the wet and icy weather.

Our sheep are now fat, woolly and pregnant. They're also exceedingly tame, especially whenever they get a sniff of a sheep nut.

I put on my usual sheep-wrangling kit: a warm fleece and a sad-looking black-and-white penguin hat that ties under my ears. I filched it from the kids' wardrobe, as it was the warmest thing I could find. The penguin has lost one of its wings where the stitching unravelled, but in the driving rain of a January morning it keeps my ears and head toasty.

Steve hands me a bag of sheep nuts that is so big that the top of the plastic comes up over my nose.

'You shake the bag of sheep feed at them, and they'll follow you up the field and into the pens,' he says.

I start trudging through the muddy field, making my usual 'sheep call': 'Sheeeeepies! Sheeeeeepies! Come on sheeeeeepieeeees!'

I've spent the last few weeks shovelling in Quality Street, pigs in blankets and home-made Christmas cake, so it's really hard going through the sticky mud.

The ewes look up, catch sight of the feed bag and belt up the field towards me, Mabel in the lead. I jog to stay ahead of them, trying to keep my feet over the soupiest, muddiest part of the grass.

The bag gets too heavy to shake and begins to dislodge my stupid penguin hat, so I heave it round to my front and clutch it to my chest.

The sheep get faster and faster, and I try to stay in front of the flock, wheezing and gasping, and batting at Mabel's nose to keep her from tearing the feed bag out of my hands.

Just before the field gate they overtake me, and I'm swept on top of their woolly white backs. The sheep always speed up as they approach the pens, and they shoot past an astonished Sunday Lunch party, who watch as, with legs like a blur and my hat on backwards, I involuntarily race past them and into the yard at the top, managing a weak wave as I whizz by.

When I finally fight my way out of the pens I have lumps of mud and sheep poo from my eyebrows down to my ankles, and I sit down on a hay bale to recuperate. For a moment I think I'm going to have a heart attack.

The foot bathing is over in a jiffy and the sheep can now go back to their field. This time Steve has the sheep nuts and I sit on the bike.

Friday, 5th January

When the weather is frosty and dry I like to go for a walk. It's my favourite type of weather. The mud and clarts freeze over and the fields look stunning when they're covered in a thin layer of white. When it's very cold everything is outlined in ice: trees, leaves and grass. The low sun and bare trees mean I can see straight through all the branches and hedges, which would normally be cloaked in green. It sounds stark, but the farm still looks beautiful.

Today Mavis and I walk down to our wood.

As a working collie, Mavis doesn't normally go on doggy-type walks, and for the first ten minutes she keeps trying to gather up the sheep. When I tell her to stop she's confused, and doesn't understand why I want to walk down through the fields when there's perfectly good ewes to herd together.

But in the forest she begins to relax, and starts to play like other dogs, sniffing and pushing her way into the thick, tangled undergrowth, chasing rabbits and bounding across the brambles in hunt for interesting smells.

Today she's like a domestic pet dog, and it's lovely to see. I sometimes forget she's only eighteen months old, little more than a pup. She races up to me, coat covered in burrs and her tongue hanging out of her mouth, and then bounces away again to investigate a foxhole she's found under a tree stump.

Saturday, 6th January

Today we've started giving extra feed to all the pregnant ewes. We need to feed them supplementary food to ensure that their placentas are secure, and that the lambs are growing properly inside them.

Each year Steve draws up a chart on the back of an old envelope detailing exactly how much poundage of sheep nuts we'll need for each ewe. We're already feeding the triplet ewes three quarters of a pound of nuts twice a day. Those ewes are inside the shed, so it's an easy job to tip the nuts into their sheep trough. The sheep carrying twins and single lambs are still outside, and the feed therefore needs to be taken out to them in the field.

Steve puts two 25 kg bags of sheep nuts on the back of the quad and pours them straight into the long line of sheep troughs in the back field. The sheep belt over the hilltop towards him, and immediately gulp down their ration of nuts.

When I'm on my own I'm not strong enough to lift a whole bag of nuts, or tall enough to pour them into the troughs. Instead, all the ewes, maddened by the thought of dinner time, run at me in a stampede, and I go over like a bowling pin and get trodden on.

Today, after I stagger in covered in mud and sheep hoof prints, Steve has taken pity and gone out to the shed to pull out the 'snacker machine'.

The snacker is a mechanical chariot on wheels that you fill with sheep nuts and pull behind your quad bike. There's a button that you press to open the hopper at the bottom of the snacker to leave little heaps of sheep nuts at predetermined intervals on the grass.

It's much more civilised, as there's no need to get off the bike, so I don't get flattened by hungry sheep and everyone gets the right amount of feed.

It also gives me time to sneak a few more nuts to my favourites, Mabel and Pudding, without anyone noticing.

Monday, 8th January

Disaster. Steve got a phone call today and has been made redundant from his part-time job as a site manager. It's not his employer's fault, but rather the result of cutbacks in their own industry. We desperately need the income to bolster the lack of profit on the farm, but finding a new job, just before lambing, and with only part-time hours, will be a nightmare.

Tuesday, 9th January

We're still reeling from the fact that Steve has lost his job, but the farm doesn't wait, and today we need to crack on with scanning our ewes.

Kevin the scanner has arrived with his ultrasound equipment to scan each of our ewes and find out how many lambs they're carrying.

Knowing how many lambs each ewe holds means that we can supply each of them with the correct amount of feed, so that in April they will give birth to healthy singles, twins or triplets. The normal gestation is roughly five months, and it's usual to scan halfway through lambing; any earlier and there'd be nothing to see. We're hoping there won't be many 'geld' or barren ewes, and that our tups have been doing their job.

We gather all the ewes into the barn and Kevin sets up his rig. It looks just like the equipment Hexham Hospital used in their antenatal department when I was pregnant with Ben and Lucy. Kevin sits in a battered old chair so he can see the computer screen. In front of him are three pots of paint: red, blue and orange.

Each ewe is coaxed down the sheep race into a small pen, where Kevin scans them and then marks their fleece with a blue blob of paint for a single lamb, a red blob for twins and an orange streak for triplets. Geld ewes get a purple spot between their shoulders.

It's hard work keeping up a steady stream of ewes for him to scan. They're all used to the sheep race and happily trot down the narrow corridor, but some stop dead when they see the scanning pen. They've never seen one before, and persuading them that we're not doing something terrible is tricky. Pudding gets into the pen but then refuses to come out even though she can see her flockmates happily standing at the far end of the enclosure.

'Come on Princess!' shouts Kevin, giving her tail a squeeze. Pudding shoots out of the crate in a huff and trots over to her friends, bleating loudly. She's scanned as twins. My favourite ewe, Mabel, is also expecting twins. Spotty Nose, the oldest ewe and leader of the flock, is carrying triplets, and Mrs Snuff is having a single.

'She might be the first to lamb,' says Kevin, waving a paintbrush at Mrs Snuff. 'Her lamb looks further on than everyone else's.'

There's one ewe that might have quadruplets. 'I can't definitely tell, but it looks like there's four in there,' says Kevin as he pushes the probe against her stomach. Sometimes a sheep 'reabsorbs' a lamb back into her womb, so there might eventually just be three. We'll have to wait and see.

There's only one geld ewe, and she gets a purple stripe between her shoulders. She might just be in the very early stages of pregnancy, so she'll stay with her flock this year and have another chance at getting in lamb next autumn.

Overall we're at 170 per cent – which works out at an average of two lambs per ewe. That's just what we want. Too many triplets isn't ideal as it means we would need to bottle-feed the third lamb, as sheep usually don't have enough milk for more than two offspring. And we don't want many singles, as the whole point of lambing is us trying to make as many lambs as possible.

Kevin packs up his stuff and races off to his next job, in Morpeth. We go home for a cup of tea, discussing how busy lambing will be this year. 'All we have to do,' says Steve, 'is keep them all alive.'

Thursday, 11th January

During his late-night check on the animals Steve spots a small, black car parked in the gateway of the back field. When he returns from putting extra hay in for the lambs it's still there, engine idling. He hops onto the quad bike and trundles down for a nearer look. As he gets close, the car suddenly pulls out of the gateway and, without putting on its lights, accelerates off down the road.

He didn't see the license plate or any of the people inside the car. Suspicious, but there's nothing to report. We're sent regular emails from FarmWatch, a rural police update, and they're reporting that there's been a spate of quad bike and equipment thefts in the area.

We sleep all night with one ear cocked, but everything seems quiet and still.

Friday, 12th January

Mavis is absolutely filthy today. She's been rolling in something disgusting and she is paws-to-ears covered in thick, black mud – very happy with herself. While the kids and I are waiting for the school bus we try to teach her to play football. She thinks we're mental. Why are we kicking a round thing at her all the time? She waves her tail at us, looking confused, then slopes off to stare at the sheep in the front field.

Saturday, 13th January

Bursting in through the door this morning, Steve drags me out on the quad to check the back field full of sheep.

'Count them! I can only see sixty-one, but there's supposed to be seventy-one!'

Mavis rounds them up and I quickly count up in multiples of two: '... fifty-eight, sixty, and one left over. There's sixty-one. Why? Where's the other ten?' I ask.

'I don't bloody know. I'll go and check over the hill,' Steve replies.

He charges off on his quad, and I recount and recheck just to make sure. Yes, sixty-one ewes, all round and fat, happily munching on the haylage in the feeder.

Steve reappears, his teeth gritted. 'Nope, there's nothing in the fields next door. I reckon they've been stolen.'

Back at home we quickly ring round our neighbours to check that they haven't suddenly found ten extra ewes in with their flock. They'd be easy to pick out, as they're Texels – big, chunky animals – and they have a blue mark on their shoulder to show that they're High House animals.

Steve's on the phone to the police and reports back on his conversation.

'Apparently we're the third case in the last fortnight,' he says with gritted teeth. 'Some scrotey bastards are taking a few sheep at a time and sticking them in the back of a transit van.'

We troop out to look at the field. We've had a few frosty nights, so there's hardly any tyre marks, but you can see that someone has driven into the field, and the ground is churned up round the gateway.

I'm gutted. We're both gutted. The ewes were in-lamb, well looked after, happy and chunky. Now they've probably been forced into some van, squashed in and slaughtered for meat. We know their ear tag numbers, but it's no use, as once butchered, who could tell where the carcasses were from?

I start muttering curses. 'I hope their balls explode and roll down the bed. I hope they all die painful and horrible deaths. I hope ...'

'Ah, it's no use,' says Steve, suddenly looking defeated. 'I should have heard them. It's never happened before, but they must have been casing out the place beforehand.'

We suddenly remember the black car from a few days ago that had been idling in a gateway late at night. It must have been them.

Sunday, 14th January

The police ring to have a chat. The coppers are very busy, so can't spare much time. The lad on the phone makes us laugh.

'Have the sheep got any identifying marks?' he asks.

'Well ... they're white and fluffy,' says Steve deadpan.

There's a cough from the other end of the phone, 'No natural markings?'

'They're Texel ewes,' explains Steve patiently, 'so every one of them is just a big, white sheep.'

We get a crime number and an instruction to contact our insurers. The police reckon that our sheep will already be slaughtered by some back-street abattoir and be in the food chain by now.

This is what the Northumbrians must have felt like 500 years ago. You spend time and money building up your flock of sheep, lavishing care on each animal, and then along comes some scrofulous border reiver who nicks your best animal for his own larder.

Fortunately, we're insured against theft, but claiming is a long, drawn-out affair, with many forms and instructions.

I can't help thinking of the poor, bonny ewes squashed into a tiny van, terrified and then slaughtered in someone's grotty allotment.

Monday, 15th January

We're feeling down about the theft, but Steve is busy gathering some lambs together to sell at two sales this week at Hexham Mart.

The withdrawal period for the scab medicine is over, so we're free to sell last year's lambs. Most of them are a good weight, as we've had them inside over November and December, and we've been feeding them on concentrates.

The fattest will go into the 'prime stock' sale. We call it the 'fat' trade, and it's for animals that are heavy enough to be sold as meat. They'll be bought at the Mart by a dealer, and shipped to a licensed UK abattoir, slaughtered and sold as prime lamb. These lambs are now almost nine months old, and they're already big, woolly, full-grown animals.

It's a difficult thing to think about, raising animals for slaughter, and I'm not trying to put a gloss on it. We have pride in the lambs we produce. We raise them to the highest standards and give them a good life. But there's no getting away from the fact that they are shipped off to be slaughtered at nine months old.

I have read and heard all the arguments from vegans and vegetarians that accuse us of animal cruelty and that no one should raise animals for them to enter the food chain. If I'd never worked on a farm, or seen how lambs are produced, I'd also find the animal production industry confusing.

But I know our lambs are killed without pain or suffering. I've seen it done. At the abattoir they're kept in special pens for a few hours before being moved through a sheep race. The sheep have done this on a farm hundreds of times before, so they don't understand that anything is different. They go through a plastic curtain one by one and are quickly killed by attaching electrodes to either side of their skulls that delivers a huge bolt of electricity and at the same time having their throats cut. It's quick and painless. There's no panic and no suffering.

And I'm proud that our meat is produced to the highest quality and to the highest standard of animal welfare.

Tuesday, 16th January

We took thirty 'fat' lambs off to the Mart and we sold them for around £85–90 each. This is a good price, especially as we're still so desperate for cash. Steve is always nervous on Mart days, as the trade depends on so many variables.

Lucy had named one of the lambs 'Thorney' and is a bit upset at the thought of her going off to be sold.

'Come on love, you're a farmer's daughter. You knew exactly what was going to happen to Thorney,' I say to her over supper.

'And she made 88 quid!' shouts Steve tactlessly from the kitchen.

Lucy sniffs, and nods, and tucks into her shepherd's pie.

Friday, 19th January

No news on the sheep theft.

I post the information on Twitter asking for retweets and get some heartfelt replies.

'Bastards. Sorry to hear this. If you need an impromptu lynch mob or witch hunt forming, happy to be in on the action.'

'They are the lowest of the low! May they catch scrofula and their "equipment" fall off!!'

'May they die roaring! Sorry to hear this. And I hear you

about the not knowing how the girls are being treated.'

I love the farming community on Twitter. Some of them have become firm friends, and are a great support even though I've never met them in real life.

From a recommendation by a farmer in Devon we've contacted a firm called TecTrace, which uses 'microdotting' technology in sheep marker paint to deter thieves. I'm going to wait to read their blurb before deciding whether to use it or not. They also provide great big signs saying 'These sheep are TecTrace protected', which might make future sheep rustlers think twice before targeting us again.

Monday, 22nd January

Steve has a job interview for a site manager. It goes very well, and he's offered the job, but only if he's prepared to work full-time hours.

'I just can't do full-time hours and run a farm,' he says to me, after eventually turning down the role. I bite my lip. If I took on more of the farm work, and had more skill on the tractor, then maybe he could do a full-time job?

We discuss it long into the night, but we keep coming up against the same problem. It's impossible to work nine-to-five in a normal job and look after the animals and do all the arable work. And I have my own marketing work to do, as well as looking after the children.

If only we could split Steve down the middle, so he could be a farmer *and* earn a full-time wage. Money is now very tight, so every day we sit down with the job sites and the paper and trawl through for a suitable part-time position.

Tuesday, 23rd January

We all have terrible colds. I have two bits of tissue paper stuck up my nose, trying to stop it dripping. Steve is complaining of a bad headache and Lucy is off school, limply draping herself over the sofa and testily demanding ice lollies for her sore throat.

No one can take any time off. It's not like the animals don't need to be fed and watered just because we've come down with a cold. I feel hot and shivery but drag myself outdoors to turn out Candy and help Steve feed hay to the outside sheep, clear water from troughs of ice and feed the indoor lambs.

When we get back to the house I'm sweating and aching. Sod doing any work. I crawl upstairs and into bed, clutching a hot-water bottle. I can hear Steve stomping around the house in search of paracetamol.

Wednesday, 24th January

Still ill. Still grumpy. Steve looks after Candy, who is his new best friend, as he gives her huge amounts of feed, even though I tell him not to give her more than a handful.

'She looks hungry,' he tells me sadly, 'so I gave her half a bucket.'

'Half a bucket?! She's not hungry, she's like a bloody ball!' I shout hoarsely down the stairs.

No reply. He must have gone out to check round the back field.

Cinders the kitten is cuddled down into my shoulder. Sumo the farm cat is stretched across my feet in bed. Everyone is emitting heat and making me feel hot and irritable. I eventually fall into a fitful asleep, until Lucy comes upstairs to say, 'I can hear you snoring from the lounge, Mummy.'

Friday, 26th January

I'm back on my feet today. But Steve has now got the lurgy and goes to bed. He never goes to bed when he's sick, so this time he must feel really poorly. He once had pleurisy and still staggered out to feed the animals in the cold until the doctor had to order him to rest. He doesn't have time to be ill. When I first met him he caught lots of bugs, one after the other, and we decided it was because he'd been so isolated at the farm so was unprepared for the onslaught of infectious viruses that I brought with me when we started living together.

He feels dreadful, so I trot up and down the stairs with paracetamol and cold drinks all day.

Sunday, 28th January

Steve is still in bed. My sympathy is wearing thin. It's hard work. I don't know how single-parent families manage.

'Stop coughing!' I shout at Steve, who is hacking away upstairs. 'You sound like a half-dead ewe!'

He hates being inside, and starts questioning me. 'Have you checked the lame lamb? Could you catch it? Grab the shears and antibiotic spray and do its feet will you? There's some amoxicillin you can give it in the cabinet.'

'Bugger off! I'm doing the best I can!'

I start to feel anxious, and make many to-do lists, trying to catch up with work in the evening and get ahead of myself.

Fortunately, Steve starts to feel a bit better, and comes down for his tea. He's still coughing and looks about 90 years old.

Thank god it's not lambing or harvest time, as we'd have to get someone in to help, which would cost money that we can't afford to pay.

Monday, 29th January

Steve is out of bed and back on his feet. He's still coughing and feeling a bit wobbly, but he's straight back out to check on the sheep.

It's almost the year-end tax return date, so Steve and I make an appointment to go and see our accountants in Hexham, clutching twelve months of farming receipts. The reception of Armstrong Watson Accountants was wall-to-wall farmers, all clutching grubby Tesco bags full of paperwork, grumpily trying to sort out their tax returns.

Gone are the days when you could 'forget' to write down what you'd bought in the stub end of your chequebook and try to claim a new car as an essential piece of farm machinery.

Our accountant Andrew is cheerful, professional and a rather good grower of dahlias. He talks through our threadbare accounts, and over a cup of coffee examines overdrawn bank statements from twelve months of farm business. After it's all finished we push our way through the crowds of checked shirts and welly boots and go and have a cup of tea in the Hexham Mart café.

Wednesday, 31st January

Today is bill-paying day, so we're crouched in the lounge sifting through the bank account to see which invoices we can pay and what we need to leave until next month. Cash is very tight again.

The mortgage payments and borrowings we made to turn our farm buildings into a brewery place a huge strain on the farm finances. Renting those buildings out to Heather and Gary for their business helps enormously, but there's no money left over. We make do with our small flock and crops, but any surplus cash goes towards fertiliser, seed, sheep feed and medicines. Every year the price of fertiliser and seed goes up. Unlike some farmers we don't have any other assets (such as cottages we can rent out) to bring in more income. We do the best we can, keep the farm going and have another job each to keep our family afloat. We're not the only ones in this situation.

In the 1970s and 1980s farming experienced a boom, and most farmers were able to make a decent income from their land. Nowadays, due to the rise in commodity prices and the failure of lamb prices to keep track, it's unusual that a small mixed or sheep farm like ours can make enough income to support a farmer and their family.

Now that we've lost Steve's part-time income how the hell are we going to cope?

The kids don't notice how strapped for cash we are. Fortunately, we have Granny and Grandad, who take them out for treats and help us with money for shoes, school uniform, after-school clubs and the occasional weekly shop.

In return, I promise them that when they're ancient and decrepit they can come and live with us.

'I don't think I'd want to,' says Mum. 'You'd make me do all the ironing.'

'I would,' says Dad. 'You can bloody well feed me pork pies and egg custard tarts in return for all this cash we're pouring into you.'

Bless them.

Friday, 2nd February

We need to start thinking about the future. If Ben and Lucy want to farm, High House isn't big enough to support ourselves as well as our children. We go and see Andrew the accountant to talk about succession. He gently suggests selling up and buying a bigger farm. It would mean a larger mortgage, but also a bigger income, and we'd be free of the debts we made when we diversified into a brewery.

It's something to think about. But we need to start now

while the children are still young enough – it'll be no good leaving it until they are grown up with families of their own.

Sunday, 4th February

It's trying to snow today. There's a few fat white flakes drifting down and lying on the farm drive. Our grass fields look tired this time of year. I still love the colour of them in winter: a sort of greeny-grey that stands out against the steel colour of the clouds. The light is flat and harsh. It's still beautiful, but in a washed-out palette compared to the lush greenness of spring and summer.

Steve is paranoid about the state of our grass at the moment. For a long time I didn't understand why he got so worked up about muddy gateways or the tractor making wheel ruts in the field. But it finally dawned on me that the grass in our fields is a crop, just as important as our wheat and barley, as the grass feeds our animals, and we need it to be in the best of health to see them through the whole year. Grass stops growing around November, and by this time of year it has lost most of its feed value. We give silage or hay to our flock to make sure they don't lose any weight and eat enough calories to keep warm in the bitter winter weather.

The only animal that doesn't need any extra feed during the winter is Candy, who looks like a furry ball all year round. Although I can tell that even she considers the grass less than appetising, as she's always desperate to get back to the stable hay net.

The grass will start growing again in late March, and during the summer months our fields are a bright emerald green.* The grass stalks grow so long that they ripple in the wind, like a huge green sea.

---

* Although of course in the summer of 2018 our grass hardly grew at all, due to the prolonged heatwave from June to August, so rather than emerald green it was a dusty yellow colour instead.

We need to look after our grass crop to make sure it remains in top condition. Steve harrows, fertilises and 'tops' the grass (cuts off the top few inches) in the early summer to ensure it stays thick and abundant. We are careful to see to it that no field gets too muddy or 'poached' by animal hooves in the wet weather, and sheep are rotated through our different fields so that no grazing becomes exhausted.

Tuesday, 6th February

Today we decided that it's time to sort the in-lamb ewes ahead of lambing. We want to put all the singles together in one field, with the larger group of twins being split between the front and back fields and all those sheep carrying triplets going into the shelter of our shed.

This means, in theory, that we can start feeding them the correct amount of sheep feed for the number of lambs they're carrying: triplets get the most, followed by twins and then singles.

The twins need to be split between the front and back fields as there's not enough grass in any single field at this time of year to make sure they have enough to eat.

This sounds easy in practice, but this morning we head out to find that all the ewes are mixed up in two big groups in the front and back fields. We need to get them all into the sheep pens and through the sheep race so that we can 'shed' (i.e. sort) them into the correct holding pens before we put them back in the field. Except we've only got two holding pens, so we'll have to put the triplets and singles together and then (once we've put the twins in the other holding pen into the field) sort them again into their correct section.

My head hurts.

There's only the two of us. I've left the children in the house, with strict instructions not to burn it down or answer the phone. It's also blowing a gale, which means we shout at each other to be heard over the roaring wind.

Off we go. I stop traffic by standing in the middle of our main road, and Mavis herds all the sheep out of the gate and up towards the farm. It's early morning, she's very keen, and she rounds them all up very fast, so the sheep hurtle towards me, while I wave my crook to try and turn them into the farm gate.

'Mavis! Settle doon!' shouts Steve.

We don't want to stress or over-exert pregnant sheep as they could lose their unborn lambs.

'Lie doon Mavis! Lie doon!'

She drops to her belly but is so keen that she can't keep still. She creeps towards the back pair of ewes, sending them skittering up the road, which panics the flock, making them run faster and faster.

I wave my arms and the sheep skid around the corner, down the front drive past the brewery and into the sheep pens.

I puff along behind, while Steve brings up the rear on the quad bike.

The first sorting goes well. Singles and triplets in one side, twins in the other.

We then spend an hour herding them out into the correct fields and sorting the triplets into the lambing shed.

Now is the turn of the mixed group of sheep in the back field. As we open the gate I notice our posh neighbour on a horse, trotting down the grassy track towards us.

'Sybil! I'm just moving sheep!' I shout to her from the gate.

'Carry on darling!' she bellows back, fag in one hand and reins in the other.

By this time Mavis has settled down a bit, and we manage to round up the ewes at a sedate trot and move them up to the gate. However, with all the shouted commands and hollering against the wind, Posh Neighbour's horse decides that it's far too noisy for comfort, and it would much prefer to be in the safety of its stable. The horse whips round and starts bucking towards the gate, nose to the grass and four feet off the ground.

Our sheep have never seen an out-of-control horse and rider before, and they burst through the gate, up the drive and into the pens in terror. Mavis can hardly keep up. Sybil manages to hang on, as her horse goes past the gate and out into the road towards home. The last we see of her she's hanging on for dear life as her horse thunders back to the stable.

'Bloody hell,' says Steve, with feeling, 'that was a bit exciting.'

We all take a moment to settle down.

Once all the sorting and re-sorting and herding has been done we put Mavis back in her warm kennel and sit down with a cup of tea and a chocolate biscuit in the lambing shed to watch the triplet ewes settle in.

The ewes are investigating their new home, nosing in the straw and busily munching on a bale of hay. Some of them already look visibly in lamb, with much wider bellies than normal.

We watch a particularly large ewe flop down for a rest: she sags back on her haunches with a groan and then bends her front legs. She rolls slightly on her side, letting her big belly rest into the straw. There's a loud fart and then she starts to chew her cud, her jaws moving from side to side as the lump of half-digested grass bulges in her throat.

I feed her a few sheep nuts and she gently takes them from my hand and then explores my face with her nose, huffing sheepy questions into my hair.

'Let's call her Bertha,' I say to Steve. 'That's a good name for a big sheep. She'll be massive by April.'

Thursday, 8th February

The local paper gets in touch about our stolen sheep. They want to do an article and take a photo of us at 2 p.m. tomorrow. We're going to look like a right couple of numpties, standing in an empty field, doing a mournful 'Where's our sheep?' face at the camera.

Friday, 9th February

Steve has had a haircut and is complaining about being almost twelve stone. I've washed my hair and am wearing my best (and only) tweed jacket, which doesn't quite meet round the middle. The photographer takes one look and positions us on a hillock in the middle of the sheep field, in an attempt, I think, to make us look less short. Steve holds his sheep crook while I lift my chin and stare off into the distance, trying to look distraught and yet noble and to avoid any chance of a double chin.

Mabel the ewe wanders up to check everyone's pockets for sheep nuts. She's very persistent and investigates the photographer's tripod. She's nose-deep in his camera bag when he tries to nudge her out of the way with one foot. Mabel doesn't like being nudged, and pushes back so that he trips over a molehill.

'Bugger off Mabel!' Steve hisses, as we plaster smiles on our faces and hold the sheep crook a bit higher. My too-tight tweed jacket flaps open and Steve has a red nose.

Fortunately, it only takes about ten minutes, and then everyone can go home. After we're back in the house I can see Mabel carefully checking and rechecking the grass for spilt sheep nuts.

Saturday, 10th February

Wedding season has begun.

I rather enjoy it, as whenever I'm working with sheep on a weekend I collect a small string of bridesmaids who follow me about, get big mud marks across their pretty pastel dresses and green sheep poo on their sparkly shoes and ask me loads of questions.

I always get asked, 'Do you eat your sheep?'

'Not all of them at once,' is the stock answer.

Candy the fat pony plays her part at weddings. For a bit of

excitement I take her out of the field and let any interested tiny wedding guests pat her while she eats the grass on the verge.

Last year I had this little lad following me around. He was from Sunderland, was called Kieran, and was wearing an amazing bright-blue three-piece suit with tie and one of those haircuts where it's dead short up the back and long and floppy on top. He must have been about seven years old.

'Can I sit on her?' he kept asking, while stroking Candy on her neck. Candy never lifts her head from the grass. You could literally put anyone on her back and as long as she has access to food she couldn't give a stuff.

'If you ask your Mum and Dad, and they come and lift you on,' I said, vaguely aware of the possibility of insurance claims due to injury from a small fat pony.

'Maaaaaaaaaaaam!!!' he screamed across the brewery. Eventually this huge bloke wandered over, also dressed in a blue three-piece suit.

'This is me Uncle Kev,' said Kieran.

Uncle Kev looked a little bit worse for wear, but duly lifted Kieran up and stuck him on Candy's back.

Kieran's face was a picture. 'I've never sat on a proper horse!' he said. He kept patting and stroking Candy. 'I can even lie down!' he said happily, lying backwards so his head rested on Candy's bum. 'It's dead comfy!' he beamed.

Candy is always on a diet due to being completely spherical, so is very soft to sit on. She didn't bat an eyelid at Kieran messing around her back, just kept hoovering up the grass.

'Can she go fast?' asked Uncle Kev.

Candy lifted her head and gave him a long, level look.

'Not really,' I said. 'Unless it's windy and I've got sheep nuts. You can get her to trot if you pull her along and someone waves a Tesco bag behind her.'

'Let's try!' said Kieran.

'Maybe not today,' I said. 'You'll need a proper hat and Uncle Kev would have to run along to hold you on.'

Even after I'd lifted Kieran down he followed me around

asking questions and visiting the sheep and the chickens. He kept asking if he could come and work on the farm. 'I'd live in the brewery, and me mam and dad could just bring me dinners.'

I wanted to take him on there and then and give him his own overalls and welly boots. He was driven away by his parents and we saw him grinning and waving in the back of the car all the way down the drive.

I hope he comes back again. I'd even wave a Tesco bag behind Candy for a bit of extra excitement.

Tuesday, 13th February

Today we went to the Mart, to sell the very last few of our prime stock lambs. They made a whopping £99 each.

Flushed with success we pop by Hubbucks agricultural store and I spot a 'KiwiCrook'. These are a new type of stick with a normal semicircular hook on one end and a foot crook and latch on the other end. It's very clever. It makes it possible to catch a sheep's foot while the latch keeps it secure so that the hoof can't slip out.

Steve decides to buy me one for Valentine's Day tomorrow. He presents it to me outside the shop and tells me happily that he's also managed to claim the VAT back on the farm account.

What an old romantic.

Thursday, 15th February

My friend Debbie is going on holiday with her husband James and her kids. I've volunteered to look after their dog, a little fluffy ginger Pomeranian called Teddy Pom-Pom.

Pom-Pom is completely unpractical on the farm as he's like a small fluff ball, and his legs are very short. This lack of leg and general fluffiness means that he comes back from walks absolutely plastered in mud.

We've all got a soft spot for Pom-Pom, even though I've seen Steve nudge him into the ditch when he spots another farmer.

Today Pom-Pom is helping us foot-bathe the sheep. By helping I mean that we've tied his long lead to the end of the sheep race. When the sheep go past him he yaps courageously at them, which encourages them along into the pens.

We shout the occasional 'Good boy!', which seems to keep him going. He sits bravely at the end of the race, eyeing up the huge ewes and running up and down at the end of his piece of string while eating the occasional bit of sheep poo. The whole shed is ankle-deep in festering mud, so by the end of the afternoon Pom-Pom is wearing a thick duvet of green muck.

He's had such a good time. I reckon he enjoys being on the farm. We towel him off and put him in the kitchen to dry. Later on I find him upside down in front of the lounge fire, fast asleep with his paws in the air.

Friday, 16th February

It's a good bonfire day, cold and dry with a good breeze. All winter we've been collecting stuff we need to burn. We're not allowed to burn plastic, sheep feed bags, dead animals (or people), silage wrap or bale nets. Everything else is fair game, if you have an exemption from the Environment Agency. Which we do.

We make a big pyre and I spend the afternoon trotting up and down, chucking on all the cardboard boxes and wrapping paper from Christmas plus all the endless bags of rubbish that seem to collect in the corners of our hemmels and barns.

I smell of smoke, but I come in happy and exhausted. If my number one favourite farm job is clearing out streams, burning stuff is my second favourite. It feels wonderful to be finally rid of all the rubbish that has silted up in the corners of the barns and sheds over the last few months.

Saturday, 17th February

Our neighbours have been inspired by yesterday's bonfire, as they've built their own on their tennis court. It's a massive pile, and their gardener has been hard at work, trekking backwards and forwards to their stables and garage, and chucking on all sorts of junk.

My dad has turned up with some spring bulbs and is pottering around my back garden, pointing out weeds and happily grubbing around in the flower borders.

Steve and I start on the morning sheep check. On our way round the front field we hear a *massive* bang. It sounds just like a cannon.

Racing back, we find Dad wandering round in small circles making 'oof, oof' noises. When he bent down to plant a daffodil bulb, a piece of metal shot past him at head height and embedded itself in our garden fence.

'I've been bloody shot at! There was a hell of a bang!' he says when he sees us.

We look at the bit of metal which seems to have come from an old tin.

We realise that our neighbours must have been throwing old cans on their bonfire, and the heat made one of them explode.

'It's like the Somme up here!' says Dad, understandably distressed that he could have been decapitated by an old Dulux paint tin.

We sit him down and give him a cup of tea and a pork pie.

Steve goes and has a quiet chat with our neighbour's gardener and asks him to check through the piles of rubbish before they're burnt.

Last year, we accidentally threw an aerosol of sheep marker on our bonfire and it went off like a bullet, shooting out horizontally and nearly taking out one of our chickens.

Sunday, 18th February

I'm digging over our allotment. We have a small vegetable patch near the sheep sheds, with a patched-up polytunnel, home-made garden shed and four vegetable beds. This year I've decided to grow leeks, potatoes and strawberries. I fork in barrowfuls of manure from Candy's stable and the occasional bag of old hops from the brewery to improve the heavy clay soil. It's a long and arduous job, and the kids jump over the fence, collecting the worms to feed to the chickens that are pecking about in the farmyard. Eventually Marjorie and Ethel join me in the allotment and watch me digging through the soil, waiting patiently as I fork over the thick clumps of weedy soil so that they can run in to snatch wriggling worms and beetles.

Monday, 19th February

It's time to hand back Pom-Pom the dog.

He gets car sick, so Debbie has rung me from the airport to tell me how to get him home without him throwing up all over my vehicle.

'Put him on your knee and open the window right down. Then, push his nose out when you go around the corners,' she says.

Suitably instructed, I get in the car, put Pom-Pom on my knee and make sure the window is wide open. It's absolutely freezing, and windy too. We set off down the road. Because my legs are so short, my seat is right up to the steering wheel, and Pom-Pom is jammed in the gap with his feet against my chest. As we go around the corner, he slides across my face and I end up wearing him as a hat.

This isn't going to work. It's probably illegal. I stop the car and readjust us both. I work out that if I stick Pom-Pom half out of the window, there's just enough room for me to turn the wheel and move the gearstick.

By the time we get to Debbie's house, Pom-Pom and I are both frozen, and wrapped in an inextricable knot with the seat belt.

Tuesday, 20th February

Dad and I have signed up to go on an 'Introduction to Lambing' course run by our local vets, in the town of Rothbury, about an hour's drive away.

I want to fill in the gaps in my lambing skills, and Dad just enjoys a day out and the chance to ask questions at people.

It takes us an hour to get there. The course is held in the town hall and hosted by a vet called Samantha. There's around ten of us there: a mix of farmers, students, smallholders and Dad.

The morning session goes through the basics of the care of the pregnant ewe, how a normal labour progresses, how to look after both newborn and the mother and, depressingly, all the different diseases she and her lamb could catch.

Then Samantha starts discussing ringwomb. Ringwomb is a terrifying condition where the cervix doesn't properly open in labour. It's a shepherd's nightmare. I ask Samantha what she recommends.

'Good question. It's a tricky one. If you catch it in time, the best thing to do is to gently push your fingers into the ewe's cervix and try and massage it, so it opens up a bit. At least enough so that you can help out the lamb.'

A few years ago, we had a ewe with ringwomb. She started the usual signs of labour: pawing at the straw, stargazing, licking her lips and pushing, but she didn't seem to progress. When we checked her, her cervix was completely closed. Normally you can fit your hand through and feel the lamb's front feet, but I couldn't even fit a fingertip through the entrance. Eventually, through patience and sheer strength, we managed to help her enough so that the lambs could be born. They were both dead. It had just been too long between the onset of labour and getting them out of the ewe. Devastating.

Everyone else chips in with their stories of difficult lambings.

Dad is enjoying himself. He's the oldest in the classroom by about forty years, and is asking many, many questions and regaling people with his adventures in lambing so far.

We all get a sandwich lunch and drive to the afternoon session, which is held in a lambing barn at a windswept farm just outside the town.

Everyone gets changed into their lambing wear, involving complicated layers, waterproof trousers and jackets, hats and gloves.

The lambing barn is huge. It holds 400 pregnant ewes and looks ultra-organised. All the pregnant ewes are corralled into three large pens in the middle of the shed. The individual lamb pens have been built around the edge of the shed, each with hay feeders and water drinkers hanging on the side of every gate. Most of the pens are already filled with ewes and pairs of lambs. This farm lambs earlier than us, and will have tupped their flock around October.

As we're waiting, two ewes are already in the middle of labour in the big pen, and one has just given birth. We want to help, but daren't step in, in case we do the wrong thing.

Where's the shepherd? Eventually we see him roaring down the hill on a quad bike, looking manic and wild-eyed. The bike screeches to a stop in a shower of gravel and he rushes into the shed and begins to pick up lambs, catch ewes and drag them into pens.

The shepherd's face falls when he spots us all awkwardly standing round the entrance to the shed. A load of clueless students is the last thing he needs in the middle of a very busy day.

The vet clears her throat. 'This is one of the busiest farms in the area. They have 800 Suffolk ewes, split into two flocks. This is the first batch, and as soon as they're all lambed, they'll be moved out, and the next flock will come in.'

The shepherd rolls his eyes. 'Aye, if the rain stops pissing it down. They're stacked three deep at the moment.' He waves a hand at all the pens. 'And there's only me, and the ould wife helping today,' he mutters.

Samantha gives a watery smile. 'Well, we'll keep out of your way, I promise.'

She leads us over to a quiet corner of the shed. We're all 'oohing' and 'ahhing' at a tiny kitchen set into the wall. There's even a hot-water boiler, and a roll-out bed! It's all very posh. This must be where the night lambers hang out.

On a table is a large, rectangular wooden box with a hinged lid and a round hole cut in one side. The vet picks up a sheep feed bag and after rummaging around in the bottom, produces a small, very dead black-and-white lamb. She pushes the dead lamb through the hole into the wooden box.

'Right,' says Samantha. 'Who's first?'

The box is a pretend ewe's uterus and we're all going to practice our lambing skills.

I volunteer. Looking into the hole, I can see a thick plastic bag which is half filled with water. She hands me a bottle of lube, and I squeeze a generous amount all over my right hand and push my hand through the hole. I can just about feel the tip of a cold, wet lamb foot.

'I've put it in a weird position,' says Samantha, with a smile. 'See if you can work out how to get it out.'

From the tip of the hoof, I work one hand down the leg and check which way the joint moves, so I can work out if it's a front or back foot. It's a back leg. I reach down the body to try and find the other leg and feel a little tail instead. I push my fingers underneath the lamb body and manage to hook a fore-finger round the second hoof to attempt to ease it out.

It feels very strange working with a dead lamb, as a ewe feels warm inside and baby lambs are often very wriggly. Finally, after a lot of squeezing and false tries, I manage to catch the second hoof, and I pull out the little cold body.

'Well done!' says Samantha. 'Remember, with a breech or backwards birth, you're going to have to get the lamb out quick, as otherwise it can drown in amniotic fluid.'

After we've all had a go with the birth box, it's time to learn how to tube-feed a lamb. For the second time, Samantha rummages around in the bottom of the feed bag and pulls out the tiniest dead lamb I've ever seen.

'Premature,' she says. 'Let's see if you can get a tube into its stomach.'

Tubing a lamb is a key skill that all shepherds need to learn. Premature or very small lambs often need a top-up of colostrum* when they're first born, so it's essential to learn how to insert a tube into their stomach so that you can give them a good feed.

I open the weeny lamb mouth and carefully push in a tube, trying to work out whether it's going into the lungs or the stomach. A tube positioned correctly will smoothly press down past the gullet and into the stomach.

'Always measure your tube against your lamb before you try and push it in, so you know exactly how far it should go,' says Samantha.

The lamb is so tiny that it takes me a few goes to get the tube past the oesophagus into the correct position.

'Great job,' says Samantha enthusiastically, as she gathers our dead lambs and drops them back into the feed bags. We all take a break and swap stories of past lambings and our farm set-ups. This is fun. I never get to work with other people, and hearing about their backgrounds and challenges is fascinating.

After a full afternoon of pretend birthing and tubing we go back into the thick of the lambing shed. It's bedlam. There's lambs dropping everywhere. The shepherd is at full tilt, rushing from ewe to ewe, picking up the newborns and coaxing sheep into individual pens. The 'ould wife' turns out to be the owner of the farm, and she's moving like a blur, injecting ewes, tubing lambs and mucking out pens.

'Here. Take this,' she says, thrusting a wriggling newborn

---

* Colostrum is the first milk that a new mother produces, full of antibodies and essential fats and vitamins.

lamb into my arms. 'Drop him in a clean pen over there, while I try and get his mother in.'

The lamb is heavy and wet with big black ears that stick out at ninety degrees from the side of his head. When I cradle him in my arms he turns his head to stare at me. I sniff the top of his forehead and inhale his baby sheepy smell. Wonderful. He goes into a clean pen lined with thick straw and starts blarting for his mother.

'They've got another three weeks of this left,' says Samantha.

The shepherd rolls his eyes at us and smiles suddenly. 'It'll be over soon,' he says. 'We're all knackered, but it'll all be over soon.'

It's been an excellent day, and when I get home I regale Steve about all the new ideas I have and how we should start organising our pens in the same style as the shed I'd seen today.

Steve humours me as I chatter on. 'We could hang up the water buckets from the side of the pen, see? Drill two holes in the side and fill them with water and then we wouldn't have to keep watering the sheep and it'd save time!'

'Let's just see how we get on,' says Steve gently. 'We've got about a third of the sheep that they had. Let's just see how it all goes.'

I go to bed, my head filled with lambing scenarios, and dream of breech births and tubing lambs most of the night.

Friday, 23rd February

When the local paper comes out on Friday I rush to find our piece on the sheep thefts. They've written a great article on how we're going to use TecTracer microdot paint and everything. But the photographs ...

I look enormous. Like a tweedy barrel. And even worse, my nose is bright red.

'Oh god,' says Steve when he sees the picture, 'we both look like a pair of grumpy fat knackers.'

'Why are my nostrils so big?' I ask him worriedly. 'Do they normally look like that?'

'I think she was taking the photograph from below,' says Steve, 'so maybe she was looking right up your hooter.'

Oh Lord. And of course, round here everyone reads the paper. No one is going to miss it.

'Eeee Sal, I saw you in the *Courant*! You looked right bloody grumpy!' shouts a passerby happily at me from across the road.

Bloody hell.

'At least it wasn't a report on you being drunk and disorderly,' says Dad comfortingly. 'Or ... being arrested for crashing your car, or being under the influence or something,' he adds. 'It could be worse.'

I suppose. And vanity aside, it *is* a good piece, and hopefully gets the message across to other farmers that there might be sheep thieves in the area.

Monday, 26th February

Today, Steve is giving me a lesson on driving the loader, so I can help more on the farm when and if he finds another job. It's not a difficult machine to operate, and I learn how to tip the forks forwards and backwards (crowding and tipping). I practise spearing a bale of hay and putting it in the sheep shed.

'Watch the roof!' shouts Steve. 'Watch it when you go backwards!'

'Watch the sheep!'

'Watch the dog!'

My spatial awareness skills aren't very well developed, and I struggle to estimate the width of the loader when driving through a narrow shed door. But there're no accidents, and I manage not to run over the kids.

Thursday, 1st March

It's World Book Day at school, when all the local parents desperately turn out our wardrobes to find some approximation of a book-themed costume.

Ben wants to go as Bear Grylls. Bear has written lots of children's books and Ben is now obsessed with the idea of Going on Adventures. He's packed a bag (plastic dinosaur, toothbrush, spare pair of pants) and drawn lots of maps with big crosses and captions like 'Tresure here look!'. We've started to lock the door in the evening just in case he decides that he wants to go on an adventure into the farm in the middle of the night.

I manage to find a pair of khaki trousers and a black top. What else does Bear wear when he's on a mission?

'A belt made of bullets,' says Ben solemnly.

I am *not* sending him into school with shotgun cartridges Sellotaped to his trouser belt. Instead we borrow a camouflage T-shirt to go with the trousers and a backpack stuffed full of Useful Things (mostly baler twine).

Lucy finds her own outfit, and decides to go to school as a Warrior Cat. When they've both left on the school bus, it takes me about an hour to tidy all the discarded outfit solutions that have been strewn around the room.

Friday, 2nd March

Steve has been offered another full-time position but is now trying to negotiate part-time or flexible hours. His prospective employers won't budge. They tell him that it's imperative that he works the full forty hours a week between the hours of 8 a.m. and 5 p.m. It's just impossible and so frustrating. After long discussion, he turns down the job. This is two full-time jobs we've had to decline. We're back to square one.

We're fed up but I find a leftover birthday voucher for a meal in a local Chinese restaurant. I don't think we've been out together since before Christmas.

During the duck pancakes Steve starts to talk about lambing.

'You realise we have twice as many sheep as last year?' he asks while shovelling in crispy duck. 'So, we'll be in the lambing shed most of the time?'

I realise this, but tentatively offer the opinion that if I'm knackered I'll be able to go and have a nap during the day. I love mid-afternoon naps.

'You'll not get the bloody chance!' says Steve, waving his fork for emphasis.

'I reckon we might have twenty-a-day lambing each day. How on earth are you going to have a lie-down? And I know your naps. You're often out for the count for over an hour. Snoring flat out.'

He starts using his knife and fork on the spring rolls. 'I just don't think you realise how much hard work it's going to be.'

'I'll be fine,' I say crossly. 'Mum can come and look after the kids for a bit, and I'll just stay in the shed all day.'

I'm sick of him going on about how much hard work lambing will be. I don't want to think about it until it happens.

I snap back, 'But what are we going to do about checking the sheep overnight? If you get another day job, *you* won't be able to do it. I'll have to get up each night.'

I am dreading this.

That evening I finally sit down and think about it, and I realise that it *is* going to be tough, and I'm going to have to get fitter and stronger to be able to cope with the day-to-day grind. I'm out of shape and struggle even to lift a single water bucket. I make a resolution to get back to the gym for the next few weeks to give me a head start. But, lying on the sofa, I realise I'm more excited than terrified.

I am going to do the whole bloody lambing. I'm going to lamb all the lambs. I might be middle-aged, and a mum, and not have a whole load of experience, but I'll show them.

I imagine myself lean and strong, with thin thighs, in attractive waterproof overalls, striding through the lambing shed like I own it.

I spend the rest of the evening searching through eBay for waterproof trousers, short leg, size 14, that don't look like a pair of plastic bags stitched together at the crotch.

Sunday, 4th March

After moaning on Twitter about the lack of women's farming wear I've been sent a link by a lady who owns a business dedicated to 'Outdoors Wear for Women'.

I flick through the website. Some of the waterproof trousers are floral. There's a pink boiler suit with leopard-skin pocket flaps. I just can't see myself in the lambing shed in a pair of pink floral overalls with leopard-skin pockets flapping at my boobs. I resign myself to wearing ill-fitting Flexothane leggings instead.

Monday, 5th March

Today I feel a bit more organised. I go to an agricultural merchant's in Hexham to buy the lambing supplies.

These include:

Colostrum tube and syringe, for tube-feeding newborn lambs who won't drink.

So-called 'titty bottles' for feeding pet lambs. In the past I used to come over all prudish and refuse to use the term, and instead insisted on asking seriously for 'lamb milk bottles and teats' with a po-faced expression. It didn't work, as everyone just asked me how many 'titty bottles' I needed.

A huge tube of birthing gel.*

Iodine to spray on newborn umbilical cord stumps after they're born.

---

* I had a brain fart halfway through lambing last year and asked for a 'large tube of KY Jelly please'. Steve was so embarrassed he had to walk to the back of the shop and look at shovels for ten minutes.

I was going to buy buckets. We always need lots and lots of buckets. But I saw the price (£4.50 a pop) and instead have decided to use ice-cream tubs, old sheep-lick cartons, horse buckets – in fact anything that can hold water and a scoop of sheep feed.

I also went a bit mad and sent off for a 'sheep-restraining device' on the Internet.

This 'Restrain-o-Sheep' device is apparently 'The easy, humane way to restrain stock. Can be used again and again for lambing, foot trimming and transportation.' It's a thick bit of plastic in a half-moon shape that fits over the top of a sheep's head, with two holes at either side for her two front feet. I can't wait to give it a go.

Wednesday, 7th March

The Restrain-o-Sheep device has arrived. I try but can't get it round any of our sheep's heads. We obviously need the extra, extra large size for our freakishly gigantic animals. After a lot of wrestling with a panicking ewe I throw it into the corner of the sheep pen.

I'll need to practice catching sheep with my new crook and my sheer athletic ability.

Friday, 9th March

Today is pen-making day. We need thirty-two separate pens made from metal and wooden hurdles fixed together with bag ties and baler twine. They'll be built around the perimeter of our lambing shed, fixed to each other and the wall.

The ewes will lamb in the main shed, and then we will coax each new mother and her babies into one of these individual pens, where they'll stay for at least twenty-four hours, to make sure they bond properly, and to ensure that the lambs can get a good feed of milk.

It's hard work lifting hurdles, and after a couple of hours I've got splinters under my thumbnail.

Finally, by about 2 p.m., we've built all the pens we need – sixteen along one side of the shed, and sixteen along the other.

The next step is to fill them all with straw and make sure there's enough water and feed buckets for each pen.

I go in search of the water buckets. I know we have them somewhere, but during the year they go walkabout.

I find a big pile of them behind a huge pile of empty fertiliser sacks. They're filthy. Rolling my eyes, I set to, washing them all in freezing cold water with a threadbare dandy brush I've found in my pony-grooming kit.

Empty sheep-lick containers make good feed buckets for sheep. I've also collected a huge pile of these, and methodically go through each one, swilling them clean with water from the trough.

Finally, I'm finished. It's taken me all day. But there's something very satisfying about the lambing shed, now that it's all bedded up with clean straw.

Steve has made a final pen at the end of the long line, and he's put in a big bale of clean straw and one of hay. It smells wonderful. I collapse back into a pile of clean straw and inspect my hands. Washing buckets in cold water has turned them red and cracked.

Cold water, cold weather and all sorts of lambing goopiness all work together to turn my hands into sandpaper and crack the skin around my nails. I make a mental note to buy a bucket of hand cream before lambing starts.

Monday, 12th March

Inspired by the facilities inside the lambing-course shed, I decide it's essential to be able to make a cup of tea when I'm on lambing duty, so I've gone and bought a big hot-water urn off eBay. It's the type of urn that you see in village halls and

WI (Women's Institute, a community-based organisation for women in the U.K.) meetings.

It arrives by courier, all shiny and chrome, and I put it on my makeshift table in the middle of the lambing barn. I've made the table out of a sheet of old plywood balanced on four car tyres. It's low to the ground and wobbles a bit, but it'll do. I dust the cobwebs off the wood and add my boxes of teabags, sugar and a pile of those little fiddly pots of UHT milk you get in hotel rooms.

I've also liberated a couple of old blackboard signs that the tearoom normally uses to write out their daily specials, and have bought a packet of chalks. I'll use this to write down what time each set of lambs were born, which pens they've gone into and whether they've had any antibiotics or had a top-up of extra milk from the bottle.

Alongside the tea urn goes a box with lambing lube, lambing ropes (to pull out lambs from inside the uterus), antibiotics, new syringes and a box of gloves.

I don't wear gloves normally. I can't find them small enough to fit my hands, so they slip off. I've had to fish them out from inside the ewe on a couple of occasions. But I keep seeing other farmers wearing gloves on TV and on the Internet, so I thought I'd better buy a box. For the look of the thing. And I do wear them sometimes to pick up the odd stray afterbirth that's left inside the pen.

My final item is a big book called A Manual of Lambing Techniques by Agnes Winter and Cicely Hill. It's a good introduction to all the different birth presentations that can happen. Well-thumbed, it's got a huge bloodstain right across the front from last year's lambing.

I think I'm sorted. I recheck all my supplies then proudly show them to Steve. He tries not to laugh. He normally does the lambing on his own with nothing more than a length of baler twine and a bottle of lube.

'Excellent, pet,' he says. 'I think you're prepared for every eventuality I can possibly think of.'

Tuesday, 13th March

The weather forecast is heavy snow. I have a friend who is an amateur meteorologist, and he gives me a call on my mobile.

'There's a huge polar vortex on its way,' he says. 'It's going to be bad. Snowdrifts and freezing winds straight from the Arctic.'

Oh god. This is the *last* thing we need, just before lambing.

At dinner time the first few flakes start to fall. The kids sit transfixed in front of the window as light snow quickly turns into a thick, driving blizzard. Within an hour or so the road is covered and the wind starts to pick up, whipping the snow into drifts.

Candy is standing by the gate with her bottom turned into the sleet and her head dipped below the stone wall. The snow has formed a thick layer on her back and neck. As soon as I open the gate she makes for her stable, without even stopping to let me put on her head collar. Once inside she tucks into her hay net. She's obviously glad to be in the warmth.

The lambing sheds aren't ready for the ewes that are still outside. We still need to move the farm machinery from inside the shed to make enough room for the rest of the flock. And we can't do that when it's snowing, as the machines will get bogged down in the field gates.

The wind starts to howl, and as darkness falls we check all the stock outside. They have plenty of haylage and shelter, so we all scurry back inside and put on the fire.

Wednesday, 14th March

We wake up to two feet of white, crisp snow. The kids are squealing in excitement, and I ring the school to tell them not to send the school bus. It'll have to be a snow day.

Lucy and Ben throw on their coats and rush outside to make snow angels and find their sledge.

Steve and I start trudging around the farm to check on all our animals.

It's −10°C (14°F), so cold that my face starts to sting. I've got a scarf so I pull that over my mouth and nose to help me breathe in the frigid air. My breath condenses on the fleece and icicles form. It's like trying to walk in the Arctic.

We feed the sheep inside, pushing out the snowdrifts that have piled up against the shed door.

The water in the sheep shed is running sluggishly, always a sign that it's about to freeze. Steve carries a clean petrol drum back to the house to fill with warm water to defrost the pipes.

Once he's finished faffing around with the water we get on the quad to feed the rest of the sheep.

The ewes are pleased to see us and run after the bike, plunging through the deep drifts. We always feed them on the rigg tops as the ground is drier. Or rather it would be, if there wasn't two feet of drifting snow.

We dig out the sheep troughs and pour in the feed. This field is exposed to the weather and the wind chill pushes the temperature down even further. I'm shivering in my two coats and ball up my fists in my pockets to try and keep them warm. I turn my back to the wind and wipe the hail and ice off my glasses.

This isn't fun. I never thought marrying a farmer would be all picnics in hayfields, but this is like role-playing Scott of the Antarctic.

'Happy Steve?' I shout over the heads of the sheep. 'Champion,' he calls back sarcastically.

Thursday, 15th March

School is cancelled for another day. It's so cold that the children only manage fifteen minutes outside before they come back in complaining of frozen feet and hands. Lucy is in tears, as she's been pushed over by the wind and her boots are full of snow.

I keep the wood-burning stove on all day and they sit, noses against the window, watching the flakes fall.

We're marooned inside the house, only going outside to defrost pipes, feed sheep and take out more silage. I can't get Candy into her field, as the gate won't open against a massive drift. She stands in her stable, her nose poking over the door.

Despite the cold, the farm looks picturesque, just like a Christmas card. The snow flattens all the bumps and hollows in the field, and the smooth white surface makes it easy to see animal tracks from hares, rabbits and pheasants.

The gutters on our farm buildings all sport dagger-like icicles, some so long that they almost reach the ground. The road and farm track is an ice rink, and just as lethal. No cars are attempting to go down our road.

We trudge round to our neighbours to check everyone has food, firewood and running water. Toni, our next-door neighbour, is an A&E nurse in Cumbria, and she's on the phone to her work, telling them that she's completely snowed in. Her children play with Lucy and Ben and make an igloo in our back garden, cutting bricks out of the snow with the help of a kitchen knife and a Tesco shopping crate.

I drive the quad bike into the back field. The rigg and furrow lie hidden under the snow, and I mistakenly drive into a big bank of ice so that the quad becomes stuck across the top of a rigg, with both wheels churning uselessly into the snow. Steve and I try rocking it backwards and forwards and the wheels spin, throwing out big clods of mud and snow. Eventually we concede defeat and decide to carry the sheep feed to the ewes on our backs. Steve can manage a whole bag, whereas I can only carry half.

On the way back walking is slightly easier, as it's more downhill, but I fall over twice, face-planting into a snowdrift. By the time we reach the field gate my face and fingers are totally frozen.

Steve sends me into the house to defrost while he starts the tractor to fetch the quad bike.

I warm up and then set off outside again – water pipes need defrosting and I want to give Steve a hand. He's pulled the quad bike into the farmyard and we put it back in the shed. At least we can use the tractor to take feed to the sheep and it won't get stuck in the bumps and hollows of the field. I wish the snow would stop, but it's still falling – big, fat flakes blown by the howling wind into huge drifts in the lee of buildings and stone walls.

Friday, 16th March

Third day of snow. Still no school. We need to go out to get some groceries and pick up a prescription for a neighbour. I ring Toni and ask if she needs any food. We're all running out of milk and bread.

A farmer has been along our road in his tractor, flattening the snow and creating a path. Steve decides to take his truck, so the kids and I pile into the car, swathed in coats and hats and gloves.

We creep along the road, avoiding deep snow drifts and icy patches. The first thing we spot is an oil tanker stuck on the corner flanked by a police 4×4 with flashing yellow lights. The tanker driver is standing next to the cab inspecting his wheels, which are jammed into the ditch on the side of the road.

'Do you need a hand?' shouts Steve out of the car window. The policeman strolls over in his luminous yellow jacket.

'Do you want me to get the tractor?' asks Steve helpfully. 'I've got a chain that would pull it out.'

'No point,' says the policeman tersely. 'We've already tried, and he's absolutely stuck.'

Promising to check back on our way home, we gingerly drive onwards. We've only gone as far as the next corner when we spot a white car in the ditch next to the road. There's a lady making hand signals at us from the driver's seat.

Jumping out, I haul myself through the thigh-high snow to reach her window.

'Sorry,' she says breathlessly. 'I pulled to the side of the road to let this car past and now I'm stuck!' She looks flustered. Peering into the car I see two young children and a very excited collie dog.

Steve offers to help and tries to rock the white car out of the snowdrift as the woman spins the wheels and revs the engine. Eventually, with three of us pushing, we manage to shove the car back onto the road.

The lady thanks us profusely and climbs back into her car.

'Right,' says Steve. 'Let's try and get to the bloody shops.'

Eventually we reach the village, get the prescription, buy the essentials and start the long trip home.

The drifts at the side of the road are huge, towering over field gates and stone walls. It's difficult to see where the road ends and the verge begins, and as we slither along we pass a council gritter and snowplough working overtime to clear the side roads.

The oil tanker has disappeared when we reach the corner before our farm, with only massive dark wheel ruts showing where a couple of tractors have dragged it out of the ditch. We finally reach home and I do the rounds of the neighbouring houses, passing out groceries and medicines. It's taken us almost all day to do a simple shopping run.

We've only been snowed in for three days, but it still shows how fast the country grinds to a halt, especially out here in the sticks.

Sunday, 18th March

The snow is still deep and it's still very cold.

We seem to be feeding most of the local wildlife as well as our farm animals. I've spotted three deer feeding at our outside hay stations. They bound off as soon as they see Mavis the dog, but judging by the jumble of deer hoofprints in the snow, they must be creeping back in the evening to feed alongside our flock.

In our sheds and stables there's a multitude of small birds trying to keep warm. I can see tiny wrens flitting in and out of miniscule cracks in the stone walls of the shed, a line of dun-brown sparrows and glossy black starlings roosting on our wooden beams and a bevy of chirpy blackbirds hunting for seed around our hay and straw bales. Strangely there are no pigeons. I wonder where they're hiding out?

Every time we feed the sheep a flock of pheasants and a family of little plump partridge appear to quickly gobble up left-over sheep nuts. I carry out some of last year's wheat seed to them and they gather around my feet when I dribble it in the snow. They must be so hungry, as they're never usually this tame.

If the snow wasn't so deep I'd walk down to the wood to see what was going on. I wonder how our badger and foxes are managing in the cold.

Thursday, 22nd March

Finally a bit of sun, and our snow starts to disappear.

I decide to walk down to our wood. The sun goes in as I pass under the branches of the beech trees at the boundary and all at once it feels wet, damp and cold. There's ice and slimy leaves underfoot and no berries on the trees, stripped by the hungry birds. I wade into one of our streams, which is very deep and clear, as the silt and sand has been swept away by the thaw waters. I can tell where the floodwaters have reached by the swathes of dead grass and branches left on the riverbank – sometimes up to a metre above the stream level.

I haul myself out of the water, slipping a little on the sticky mud, and then take a swift walk back along the stubble field, with clods of grey clay clinging to the soles of my welly boots. I have a quick stop to examine some animal prints that look very like the track of a badger – five pad prints with claw marks clearly visible in the soft soil by the field edge. I hope very much that our sett has survived in the Arctic weather.

I wash the mud from my boots in the deep pool below the ash tree at the perimeter of the wood. I've never seen the water so clear and cold. There's no sign of the tiny fish I often spot in the summer. Instead plenty of white snail shells are dotted around the mud and on top of the slippy rocks.

Saturday, 24th March

The floodwaters have gone and it's much drier today, so we're planning to move all the pregnant ewes from the fields outside to join their flockmates inside the lambing shed.

Mavis helps us gather the sheep in the field and moves them towards the gate. They trot smartly up the lane and pour into the building that will be their home for the next few weeks. One ewe even manages a bounce and a kick in the air.

Most of them are very visibly pregnant. Many of them have that unmistakeable pre-lambing shape, when you can see the lambs lying low down in their bellies on either side of their spine. One or two of the sheep are already 'bagging up', when their udders start to fill with milk, which is a sign that they are going to give birth very soon.

Bertha the ewe is still huge, but Hatty the ewe is absolutely gigantic. She's a big sheep anyway, and she's been scanned as triplets. She waddles when she walks, and is still around four weeks off her due date. I've been sneaking her the odd extra sheep nut, and now she's as friendly as a pet dog. She follows me around the shed, panting a bit with her extra weight, and sticking her nose hopefully into my trouser pocket.

'Stop feeding her extra,' says Steve when he realises what I'm doing. 'She's so big anyway that she'll explode before she's ready to lamb!'

Hatty hopefully inspects every bucket and investigates each pocket for sheep nuts. She's got a lovely smiling face, a bristly chin and marvellous fluffy black ears that stick out hor-izontally from her head. I sit down in the straw with her and she sinks down on her belly, poking one leg out to get comfortable. She makes a heartfelt sigh.

'Please don't explode,' I whisper softly to her and scratch the underside of her neck. Her ears flop out and she starts to chew her cud. I could honestly sit here all evening, chatting to Hatty and relaxing in the straw.

SPRING

In spring, the weather (hopefully) starts to pick up just as our lambing season begins. Lambing lasts all through April and May, and is a crucial part of the year for us; our farm profit relies on us producing lots of healthy lambs that, once weaned and fully grown, will then be sold in the autumn. It's also time to start drilling the rest of our crops and spreading fertiliser on those already in the ground.

Sunday, 25th March

Steve is drilling a field with barley. The seed comes in huge bags and is bright red. He pours it into the drill, calibrates the machine so that it will be planted at the correct level, and sets off into the front thirteen acres. I ride with him in the girlfriend seat for a couple of rotations around the fields. It's a bit faster than ploughing, but there seems to be much more fiddling around with the onboard computer to ensure that the seeds are drilled into the ground at the correct depth and number.

I climb back down and wander off to do my own thing. At lunch time my phone pings with a forlorn text message asking for a cup of tea and a sandwich. I can feel generations of farmers' wives frowning as I wait at the top of the field for the tractor and then throw up a meagre lunch of a sausage roll, a can of Coke and a Crunchie. Steve nods his thanks and slams the door, already moving in a wide circle back into the drill line. He'll drive with one hand and eat with the other.

By the end of the day he's managed to get all the barley into the ground. It's a huge relief. The only bits he has left to roll today are the headlands, which are the sections at the top of each field. It only takes half a day and then he's down from the tractor, and in a great mood. It seems to be a constant rush or struggle with the weather, but we've had dry days for a day or two, which has made a huge difference. Now all we can do is wait for the crops to ripen.

Tuesday, 27th March

No one has started lambing yet. There's still nothing happening. I sit in the shed and stare at the ewes, and they all calmly stare back. Most of them are lying down and happily chewing their cud. It's fascinating and almost hypnotic to watch. At first the sheep makes a few convulsive swallows, and you can clearly see a lump of half-digested grass travel up her throat until it pops into her mouth. Then her lower jaw starts to grind, round and round in circles, as she crushes the grass into a thick paste, which is swallowed again with a look of benign contentment.

Some of the ewes carrying triplets are so big that they find it uncomfortable to lie down. They shift their weight from side to side, grunting. Mrs Snuff is so large that she's resorted to sitting like a dog, with her front feet stuck out in front, propping herself into an upright position.

I have a lot of sympathy. When I was pregnant with my babies, I remember that feeling well. I had so little room at the end that I had constant indigestion and just couldn't get comfortable to go to sleep.

There's a lot of snoring and farting. The ewes spend much of the day fast asleep, conserving their energy.

Wednesday, 28th March

Still no lambs. I'm getting twitchy and impatient.

I've just bought a new box for all our medical supplies, such as antibiotics and painkillers. I spend my time organising them. New 'sharps' or needles in one side, bottles of medicine in the other. I make many cups of tea with my new urn, leaf through my lambing book and find an old garden chair to sit on while I scan the sheep.

'Don't watch them,' says Steve. 'You'll put them off. It's like boiling a kettle. If you watch it, it takes ages.'

I've taken to lurking around the shed, and nonchalantly watching the sheep out of the corner of my eye to see if anyone's labour has started.

Thursday, 29th March

Still nothing lambing.

When we fed everyone this morning, one ewe had droopy ears and wouldn't come to the trough.

'Aw man, it looks like twin lamb,' says Steve, climbing over the gate to get to his medical supplies.

Twin lamb is another name for pregnancy toxaemia. When a ewe is carrying twins or triplets and she's not managing to eat enough feed, she starts to draw on her own fat reserves. This releases ketones that can poison the ewe and her lambs. Sometimes lambs are born blind.

There's no definite cure, but we rely on injecting the ewe quickly with warmed-up calcium borogluconate under the skin, plus a shot of multivitamins. Our ewe also gets a drench (drink) of sugar syrup and electrolytes to try and increase her energy levels.

They don't always recover, but since we've caught this mother early, we're hopeful that she might pick up.

Friday, 30th March

The ewe is looking a bit better. We keep dosing and drenching her to make sure she improves. We're desperate not to lose any sheep, especially at this stage of lambing.

This evening she's sitting up in a hospital pen munching on haylage with her ears flicking backwards and forwards. It looks like she'll recover now, but we're keeping an eye on her, just to make sure.

Saturday, 31st March

Finally! Last night about 11 p.m., Mrs Snuff went into labour. Her lambs are a little bit early, as we didn't expect her to go into labour for another day or so.

She delivered all three lambs by herself, and they are tiny black-and-brown-speckledy triplets. Mrs Snuff is a Suffolk cross (so she has black legs and a black head) and her lambs have lovely big brown and black splotches across their backs and stomachs.

We quickly get her into a pen and string up a heat lamp for the lambs. This will keep them warm and toasty. Steve then tube-feeds all three with artificial colostrum milk. It'll just help to keep their stomachs full and make them strong enough to feed themselves. None were standing when we left last night, but this morning two of the lambs were up and about and nosing under Mrs Snuff for her milk. The littlest was warm but hadn't stood up yet, so we gave him another tummy full of colostrum, just to try and give him a bit of strength.

Mrs Snuff is a great mum. She's very protective and has licked each lamb clean from nose to tail. She stamps her foot to warn Sumo the farm cat to keep away. Sumo is on the hunt for afterbirth, which he drags away to feast on in another corner of the barn. Disgusting ...

Monday, 2nd April

I'm checking the ewes every morning at 6 a.m., then every 2–3 hours until 9 p.m. Steve does the night shift from 9 p.m. onwards. He still hasn't found another job and we're horribly skint, but at least it means I'm not doing the entire lambing on my own.

Lambing inside a shed has many benefits: it's warmer for the ewes and it's easier to check the flock and spot if a mother needs help in giving birth. We used to lamb the entire flock out in the fields, but Steve found the late-night shifts much harder, and the ewes would wander off to the edges of the field to give birth, and in the dark we often didn't spot if they needed help.

Lambing inside does have one major problem, however.

When she's about to give birth, an expectant ewe will remove herself from the flock and go to a quieter place, further away from her mates. She can't do this in a shed, and although our sheep have plenty of space, this lack of 'separation' from the rest of the pregnant ewes means that she's more at risk of other ewes 'pinching' her lambs. The sheep are naturally jammed full of oxytocin hormone at this point, and it sometimes makes them believe they've given birth when they haven't.

One ewe will give birth, and another ewe will nip in and start mothering and licking her newly born lamb. These 'false mothers' convince themselves that this is their lamb, and they become very attached. Once another ewe has licked a lamb, it's very difficult to persuade the original mother to allow her baby to feed, as it now smells of another sheep.

To prevent this, we check the sheep every 2–3 hours around the clock, to try and make sure that no babies and mothers get mixed up.

Of course, the other side of the coin is that some ewes will give birth, refuse to lick or clean her lambs, and instead wander off to the hay rack for a spot of lunch, leaving a pathetic lamb shivering in the straw. When this has happened in the past I've had to search through all the flock to find the miscreant ewe and reintroduce her to her baby by penning them up together. Sometimes this works, and sometimes it doesn't.

Tuesday, 3rd April

Mrs Snuff is still the only ewe that has given birth so far, and she's fast asleep in her pen, with all three lambs curled up next to her.

We have sprayed iodine onto the stubs of their umbilical cords to prevent against common infections.

Sheep have two teats to their udders, and usually make only

enough milk for two lambs. When triplets are born we need to take the smallest away from the mother within forty-eight hours and feed them on the bottle with powdered lamb milk.

This sounds cruel, but the third lamb wouldn't have a chance of thriving without being artificially fed. I've actually seen a ewe lie down and suffocate a third lamb rather than look after it herself. She knows how many she can mother properly. Sheep can count, it seems!

We reach down into the pen and weigh up which will be our first 'pet lamb'. The smallest of the three has little black ears and four little black feet. I pick him up and tuck him into my jacket. He starts to bleat and struggle, but Mrs Snuff doesn't even wake up.

I carry him over to another shed, where I've prepared a pen full of straw, with a heat lamp suspended a few inches above the bedding. Sitting in the straw, I wedge him on my knee, pick up the warm bottle of milk and try to slide the teat into his mouth.

He fights and wriggles, so that the milk drips out of his mouth onto his chest. He's not hungry enough yet, so I give him a stroke and put him down into the straw. I'll try again in an hour or so. He immediately starts blundering around the pen, blarting for his mother.

Pet lambs are very cute to begin with, especially when they're first born. Lucy and Ben come out to see the lamb after school and give him the name 'Fred'. Fred lies curled up underneath the heat lamp with his black nose resting on his tiny hooves. Lucy and Ben coo over him and want to give him cuddles. I remind them that by the end of lambing 'cuteness fatigue' will have set in, and the whole family will be heartily sick of pet lambs and preparing milk bottles.

Wednesday, 4th April

I've bought some very unflattering Flexothane lambing leggings. I hate boiler suits, as they don't fit round my magnificent monobosom, and waterproof trousers make a rustling noise and give me crotch sweat. But the leggings are so long. They have braces that are supposed to clip onto your waistband to keep them up. I'm going to have to sling them round my neck to stop them dragging on the ground. Attractive. Yet strangely comfy.

I feel prepared. My tea urn is now switched on twenty-four hours a day. I'd originally filled it from the sheep trough, but a medical friend pointed out that it was a sure way to get a nasty bacterial infection called clostridium, so I poured it out and filled it again from the water tap.

No more lambs yet.

Mrs Snuff is doing well, and her lambs are bouncing around the pen. We move her into a bigger area of the shed where she can stretch her legs, and her lambs can strengthen up by running races around the straw.

Fred the pet lamb is now gulping down four bottles of milk a day, and smearing lamb poo all over my wellies and trousers. As soon as he sees me enter the shed he races around his pen blarting furiously, and tries to climb on my lap to get to the bottle. When he finally starts to suck, his little black tail wiggles furiously from side to side. He'll happily fall asleep on my knee and loves a good neck scratch.

Thursday, 5th April

At the 6 a.m. check I go to feed the ewes as normal, and hear the tiniest 'meep, meep' noise from the middle of the flock.

A hogg* ewe has popped out a single lamb on her own. She's

---

* A hogg is a one-year-old first-time mum that hasn't yet been sheared; they tend to just have a single lamb as their first baby.

already cleaned it up, and it's happily wobbling round the shed following its mum closely.

The kids have named it Abby. We mark all single lambs with capital letters – A, B, C, etc., and the pairs/triplets with numbers, which are marked on the mother as well. It helps us keep track of whose lambs are whose. Although when I'm very tired, I sometimes forget what number I'm up to …

Newborn lambs are the epitome of cuteness. Abby has knobbly knees, fuzzy legs and a compact little body with close-curled white wool. Our Texel/Mule cross lambs always seem to have huge ears that stick out sideways to their heads. Some of them have lots of wrinkly skin, as if they're wearing a jumper that's too big. Eventually they 'grow into' this skin and it all smooths out. Some have black splotches on their faces and legs, but Abby is pure white, with very long coal-black eyelashes and a high forehead crowned with a tuft of downy wool. She blarts for her mother with a high-pitched baa, and the hogg rushes over and starts cleaning her from nose to tail with her big, rough tongue.

Friday, 6th April

Lambing has now properly kicked off. All goes well until after lunch, when I need to help a ewe give birth to her second lamb. She's already had a nice, healthy single, but the second lamb is stuck. There were a pair of lamb hooves just poking out and she was straining and pushing, but nothing else was happening.

It's a weird feeling reaching inside a sheep. Sometimes I accidentally stick my finger into the unborn lamb's mouth and feel a row of tiny sharp teeth and a flickering tongue or touch a real jumble of legs and need to sort out which legs belong to which body and whether they're at the front or the back. Sometimes when you catch hold of a lamb's feet inside the ewe the baby jerks them back out of your fingers. It's a lovely feeling as you know it's alive and just minutes away from being born.

I catch hold of the front two legs and gently pull alternately on each one, so that the lamb starts to slide out of the birth canal. I see a tiny pair of nostrils and stretch each front leg so that they straighten out, and smoothly tug downwards. The lamb slides out of the ewe and flops onto the straw with a gush of amniotic fluid. The baby must have been caught on the shoulders.

I pull all the birth sac away from the baby's nostrils and eyes and watch it shake its head, with tiny ears flapping. All healthy lambs seem to do this when they're first born. The baby is wet and steaming in the cold air. I pick him up and bring him around to the ewe's face, and she immediately starts licking and 'chuckling' at him. It's a comforting sound that all new sheep mothers make and I love hearing it.

The ewe is marked as carrying triplets, so I have another feel around inside to see if there are any more babies. I can just touch one hoof as the baby is lying deep down inside the ewe, so decide to leave her to lamb herself. I move them all into a pen, and thirty minutes later there's a smug-looking ewe and a third lamb lying on the straw. Clever girl. She's cleaned them all up and they are beginning to stand on wobbly legs. Time for a cup of tea from the urn.

I'm getting more and more experienced at telling when a ewe is in labour. Sometimes the signs are obvious. A labouring mother will paw at the straw and get up and down, moving around in discomfort. You can see them stargazing, with their noses pointed straight up into the air, and all of them start to strain and push when the contractions hit. Some ewes lick and chew or grind their teeth. Then you should see a 'water bag', the birth sac, filled with amniotic fluid hanging down from the ewe's vulva. Then two little feet sticking straight out and maybe a nose. After about thirty minutes the lamb should be hitting the straw, and the ewe, maybe after a short rest, should start cleaning and licking and chuckling to her baby.

The kids are in full helping mode. They're mucking out pens, filling water buckets, feeding sheep and moving mothers and their lambs into different pens. The pace of lambing is picking up. We're going into the busiest period, when all the flock seem to want to lamb at the same time.

The weather is still filthy, hissing down with rain and freezing cold. It'll become a problem when we need to start moving ewes and lambs outside to make more room in the lambing pens for new mothers.

Saturday, 7th April

Today was a typical lambing day.

Steve checks the lambing sheds at 9 p.m. and 11 p.m. (and usually lambs a few as well at that time of night). He's then up at 2 a.m. to do the middle-of-the-night 'look round', when hopefully everything is nice and quiet.

I get up blearily at 5 a.m. and stagger out to the lambing shed in my pyjama bottoms and no bra, with my boobs stuffed into my warmest jacket. I stumble around the pens praying to the Goddess of Lambing that nothing has started in labour. I can't help thinking that Steve's 2 a.m. morning check disturbs them all, as when I arrive there are usually a few babies to greet me.

I go back to the house to get dressed properly and drink my first cup of tea. I'm basically surviving on tea, Diet Coke and Mars bars. I get the kids out of bed, dressed and fed and leave them to play or in front of the TV.

I do most of the day work to give Steve a chance to catch up on sleep. Although he's usually too anxious about the lambing to relax, so tends to give me a hand until at least the middle of the afternoon.

Then it's quickly back into the shed to be greeted by at least one pair of newborn lambs, which need to be coaxed into a pen with their mother.

I give the ewes their morning feed and everyone a clean bucket of water. While I'm doing this I'm checking each new mother for problems, splashing out iodine on umbilical cords, and keeping an eye on any off-colour lambs. Then off to check the pet lambs and feed them all a bottle of milk. Fred still needs a bit of help to drink from the bottle, so I sit in the straw and give him another lesson, trying to think patient thoughts.

Once everyone has finished their breakfast Steve does the rounds of the sheep outside. He buzzes about on the quad bike, clutching the heated handlebars as it's freezing (and raining) out there.

Another quick cup of tea and it's back to the shed to start moving healthy mothers and babies into their nursery paddock, and we both start mucking out pens. They need to be cleaned down to the concrete to avoid infection, so after removing all the dirty straw I chuck a disinfectant powder on the bare floor and wait for it to dry.

Time for another cup of tea. Is it my third or fourth? Not sure.

I'm back out again to see a ewe in labour. This time it's triplets, and she manages well until the third lamb. It's huge and gets stuck at the shoulders, so Steve gives her a hand while I lie across her neck to keep her still. Once all is born it's into a clean pen with more iodine and an injection of antibiotics for the ewe, as she's a bit sore after her ordeal. I rig up a heat lamp to give the lambs (who are looking rather dazed) a bit of warmth.

And then more feeding and watering, and I bed up the clean pens ready for the afternoon push. More checking of pet lambs, and lots of moving about of lambs to the nursery shed, and then into the paddocks. Some new lambs need to be castrated with tiny rubber rings, and one of the newest pet lambs needs to be taught how to suckle.

It's still chucking it down, and we're running out of space for new mothers and babies, as we can't turn out the slightly older ewes and lambs into the icy rain. They'd freeze to death. So every spare inch of shed space is jammed full of sheep as we try desperately to keep them all warm.

It just goes on and on and on. I barely see the children,

unless they come out to help. Mum is an absolute godsend, and keeps a steady stream of hot dinners, lunches and cups of tea on the go.

Last check for me is 9 p.m. Praying that no ewe has decided to lamb, I scan the shed and see a sheep at the back in the familiar stargazing position. Bugger. I can see a dark bag of liquid hanging from her hindquarters, so I switch on my tea urn and settle down to wait. After around half an hour and a lot of straining and groaning she pushes out a nice, big single. I swoop into the straw, pick him up by his front feet and walk backwards into a clean pen. The mother, with afterbirth slithering behind her, totters after and immediately starts her mothering chuckle, while carefully licking her lamb. I slosh iodine on the baby's umbilical stump and check everyone else.

Back at home I peel off waterproof trousers and poo-caked boots. Mum wrinkles her nose at the smell. Every pair of trousers or coat in the hall is permeated by the strong smell of lanolin and wet wool. I flop down into the chair and drink another cup of tea. Bed. Lucy waves vaguely from her bedroom as, zombie-like, I slide into bed, and fall almost instantly asleep.

Monday, 9th April

The kids are now off school. Lucy throws herself into helping. She's old enough now to fetch buckets, bed up pens and feed ewes, and she's a great help with the pet lambs. I often find her snuggled into the straw under the heat lamp, covered in a snoring pile of different-sized lambs. She also has the most patience in teaching dim lambs how to suck on the machine.

Ben is different. He's still only little, and his favourite thing to do is ride his bike round and round the brewery car park and make up long, complicated stories about imaginary ninjas.

I love them both so much and it's difficult over the Easter holidays, as they see their friends going on holiday or enjoying exciting days out, while we're all stuck at home, tied to the farm.

'It's all relative,' says Steve that evening. 'Some kids would kill for the freedom of a farm, and the chance to look after all the animals.' I can see that, but I also want to make sure that they get the opportunity to try other things, and to see all the stuff their friends get to see.

On the other hand, the kids are discovering the benefits of having my parents living with us for a bit.

'All my clothes are flat!' exclaims Ben one morning, when he looks in his wardrobe. I hate ironing and never do it, but Granny irons everything, including dishtowels and socks.

'They're not flat,' says Lucy witheringly. 'It's just that Granny has ironed our clothes cos Mum won't do it.'

Grandad takes the children out on adventures and bike rides, and down to the beach for fish and chips.

'You'll have to last forever,' I say to them over beef stew that evening. 'I need you both lots, and so do the kids.'

'Well I don't exactly plan to pop my clogs just yet,' says Dad huffily. Mum gives another helping of stew to Steve, who winks at me through a mouthful of home-cooked food. He is also getting used to the joys of regular meals and ironed clothes.

Tuesday, 10th April

My greatest horror is having to lamb either Bertha or Hatty myself. They're now the size and width of coffee tables. It'll be OK if I can lure them into a pen, as I could at least then pin them against the fence and reach round to help them lamb if need be.

I've always been taught that the proper way to move a ewe and her lamb into a pen is to pick up the slimy baby by its front legs, make a squeaky 'meep-meep' lamb noise in the style of the cartoon Roadrunner and wait for the mother to follow you. Sometimes they do, and sometimes they don't.

Today I catch a ewe that needs a hand, but she has such a

long body, and my arms are so short, that I can't pin her against the wall and reach round to her back end to pull out the lamb. I have to manoeuvre her into a pen, almost slipping a disc in the process, and then wedge her into one corner with my knee to hoick out the lamb. I'm sweaty, bright red and covered in goopy amniotic fluid.

I wish I was a bit taller. It would make lambing so much easier. 'It's not your height, pet,' says Steve, 'it's because your arms are freakishly short.'

Thursday, 12th April

We're now in the thick of it. I have mothers and lambs in pens everywhere. The rain is still falling and we're running out of space to put everyone. I haven't eaten a hot meal for a few days, and I'm managing on about three hours of sleep. Steve has cut his thumb slicing open a straw bale with a rusty knife. I don't think I've brushed my hair since Monday.

We have six pet lambs so far, including Fred and Fuzzy. Poor Fuzzy. His mum didn't want him and refused to let him suckle, so we took him off her and made him into a pet lamb.

He refuses (or doesn't understand how) to suck on the bottle, he's got a bit of scour (diarrhoea) and he smells very bad. With his lack of personal hygiene, and being a bit underweight, he's a sad specimen.

Steve hasn't got a lot of patience with pet lambs. He gives them a couple of tries on the bottle, and will stomach-tube them milk for a few days, but he doesn't have time to sit with them to coax them to suck on a teat.

Therefore, every few hours, I sit with Fuzzy (and his smell) in the pet lamb pen and try to persuade him to drink. He won't open his mouth, so I carefully prise his teeth apart and gently push in the teat of the milk bottle. He chews it and the milk spurts out the side of his mouth, down his chest (making him even more stinky), but some does go down his throat.

Today he even manages to suck about 50 ml of warm milk

himself, but then loses the knack and starts to chew and lick rather than suck. If you gently press on his nose, it sometimes makes him remember the sucking 'feeling' on a bottle. Also, scratching him just above his tail can help to encourage his tail to wiggle and him to suck. Although I still get covered in milk, and Fuzzy seems confused. He has the hunched back and baggy skin of a poorly lamb, and he shakes and shivers, even when he's lying under the heat lamp. I'm tempted to knit him a jumper ...

He just wants his mum – who is having a high old time out in the paddock with her single preferred offspring.

Saturday, 14th April

Along with Fred and Fuzzy we now have nine pet lambs. Mornings are a blur of bottle-washing plus measuring and mixing out lamb milk. It's starting to take up too much of our time, so today we've decided to break out the Titty Machine.

That's not it's actual name. It's called the Milk Maid 2000 and it was bought by Steve's parents about twenty years ago. It cost a fortune in its day, and is marvellous, but also very temperamental and has to be treated with caution and respect.

The Titty Machine is the size and shape of a fridge and holds a hopper of milk powder, which is mixed automatically with water and passed through a heater to produce a flow of tepid milk through two thin lines, which can be attached to a pair of rubber teats.

It needs to be thoroughly cleaned every day to avoid the lambs getting an infection and coming down with scour. Lambs aren't usually the brightest buttons, and a lot of them need one-to-one training on how to suck on the machine. Some like to chew on the teats, or in Fuzzy's case, suck on the screws that hold them firmly onto the side of the pen. Otherwise, all they really want is warmth from the heat lamps, company, warm milk and space to bounce in-between feeds.

We haul the Titty Machine out from under a load of pallets

and set it up next to the lamb pen. It needs a thorough clean, so we march backwards and forwards with buckets of hot water and Fairy liquid until everything is clean and sparkling. Connecting it up to the power supply is always a tense moment. Thank god, this time it hums into life, and Steve and I take turns squatting in the pen with each lamb, showing it where the teats are, and how to suck them. Fuzzy needs a lot of persuading.

But that's a good job done, as it means I don't need to wash endless bottles morning and evening. Especially as my hands are beginning to develop deep cracks and fissures and I'm slathering on gallons of E45.

Monday, 16th April

I'm covered in iodine stains, have unidentified goopy bits hanging off my sleeves and am wearing the usual terrible selection of hairy coats and hats to keep warm. Sometimes two coats at once, so I look as wide as I am tall. The weather is unspeakable. Icy sleet whips past me when I open the lambing shed doors, and the ground is ankle-deep in mud.

Today we've had a set of quadruplets. This is quite unusual, only the second set of quadruplets that I've seen in my ten years on High House Farm. They're all a good size, and nice and chunky, with black noses and speckled faces.

They're very cute and I'm very excited. I call the kids over to see.

'Great,' says Ben, munching on a packet of salt-and-vinegar crisps. He's not very enthusiastic about lambing, as he's seen it all before, and I've taken him away from his favourite TV programme (*Teen Titans Go!*). He sits on the quad bike peering into the lambing shed. Lucy is more animated, and helps me iodine all the lambs while giving them all a stroke. I take a few photos of the quadruplets and send them to my friends on Facebook and Twitter.

This afternoon we have a case of ringwomb.

I investigate inside the ewe and can feel a cervix that is only open one or two centimetres. The sac full of amniotic fluid is bulging through the tiny gap, and as I move a finger into the cervix entrance it bursts, gushing fluid over my hand, arm and front of my jacket. Lucy leans all her weight onto the sheep's neck to keep her still, and I kneel behind her and keep my fingers slowly massaging and stretching the cervix entrance.

After around twenty minutes I manage to get the nose and one hoof out of the cervix. I can't find the other hoof inside the womb, no matter how carefully I follow the leg back from the shoulder. It must be tucked right down beside the body. I feel the lamb's tongue flick over my fingers as I painstakingly wrap one forefinger round the tiny foot and pull. If I can't find the other front foot, I'm going to have to try and pull the baby out with one leg. The ewe has been labouring now for over an hour, and she's tiring.

I'm kneeling behind her, so I sit back on my haunches and start to tug. The sheep makes some loud groaning noises, and I hate to hear it, but I daren't wait. I put all my weight into it, and thankfully the baby starts to slide forward. I can now hook two fingers round its front leg and keep pulling out and down until the cervix gives and the lamb slides out.

The baby isn't breathing, so I strip off the birth sac and quickly rub down the lamb's face, squeezing all the gunk out of its nose and mouth. I grab a big handful of straw and furiously rub at its chest and stomach, mimicking what a mother sheep would do with her rough tongue. There's still no sign of a breath, so picking up a long piece of straw, I try gently tickling the lamb's nostrils to see if I can make it sneeze.

The baby does a tiny sneeze and its chest gives a heave. The first breath. But there's no following second breath, so I leap to my feet, grab the baby's back legs and start to swing it, head down, backwards and forwards in large arcs across the straw. Birth fluid trickles out of its nose and mouth and spatters the bedding. It gives a convulsive wriggle in my hands and I lay it back down under its mother's nose and keep on rubbing and squeezing its chest and stomach. It takes another breath. And then another. And lies there on its side in the straw with its flanks going up and down like a tiny pair of bellows. The mother, exhausted, reaches out her muzzle and starts nosing down the little body. A faint chuckle emerges from her throat.

I flop down into the straw beside the lamb. The baby still looks weak, but there doesn't seem to be any damage to the front leg or shoulder that was trapped inside the ewe.

I rig up another heat lamp and carry the lamb into the pen. Making my 'meep-meep' sounds I coax the stiff mother onto her feet and she lies thankfully down in the clean straw, and starts to chuckle and clean her baby.

I give her a big injection of painkillers and antibiotics and cross my fingers that both mother and baby survive. It was a difficult birth, and they're both sore and bruised.

Tuesday, 17th April

The ringwomb mother and her baby are both doing well. They don't seem any worse for yesterday's dramatic scenes, and the lamb is now bobbling around the pen, butting the ewe to get at her milk.

Bertha and Hatty have concocted a great trick. At feeding time they both now stand right in the middle of the trough, refusing to move an inch, and I trip over them both, drop the sheep nut sack, and the two big fatties quickly hoover up as many pellets as they can fit in their mouths.

They better produce three ginormous lambs each, as otherwise I'm not sure where the feed is going …

Thursday, 19th April

First thing this morning I can hear the pet lambs bleating in a higher pitch than normal. They sound hungry. Their pen is an old stable, and as I walk towards it I see a thin stream of white liquid trickling out from underneath the stable door.

I pull open the door. The straw in the pen is wet and squelchy, and the lambs are all huddled in one corner. The Titty Machine is lopsided, with both teats hanging off the front. It's pouring out a stream of milk, soaking the lambs, the straw and the concrete floor.

Investigating further I realise that the lambs have chewed through the milk lines, sending the machine into overdrive and pumping out gallons of expensive lamb milk.

I lunge for the 'off' button and the machine shudders to a halt. Waves of milk lap at my welly boots. I want to cry. Lamb milk powder is very expensive, and the machine has mixed together and wasted a whole bag.

The pet lambs cluster round my ankles blarting loudly and demanding to be fed. I pick up Fuzzy and cuddle him. He smells even worse now that he's been paddling in curdled milk all night. I push them all into the corner of a pen, put a hurdle in front and start to drag out the wet straw. It absolutely stinks after a night under the heat lamps. It takes me a good hour to muck out the pen, wash down the concrete with clean water and bed up again with clean, dry straw. The pet lambs watch my every move. Luckily I have a couple of new milk lines, so I swap the chewed ones over, open a new bag of milk powder and attach some new teats.

The machine starts humming again, and after I remove the hurdle the lambs lunge in to start sucking away, their little tails flailing from side to side.

This time, to avoid a repeat performance I tie the milk lines to the hurdle higher up, where the lambs can't reach them. Like most babies, they chew everything at this stage. They're very curious, and like to investigate anything new or unusual with their mouths. And their teeth are sharp!

When I'm in the pen with them they mouth at my welly boots or chew on my coat or zip. When I'm sitting, some even rear up on their hind feet, plant their front hooves on my knees and have a good chew of my hair. I've learnt to tie it up out of harm's way to stop them swallowing big hanks of my ponytail.

Finally, all is done. I go back into the lambing shed and make myself a cup of tea from the urn. You need reserves of patience in this job, to stop yourself taking disappointments out on the animals, or just giving up and walking away. It wasn't the lambs' fault that they damaged the machine – it was mine. I should have realised that the milk lines were just the right height for curious mouths. Lesson learned.

Friday, 20th April

Today I'm existing on sausage rolls, chocolate hobnobs, Diet Coke and plenty of coffee. Steve has gone grey through lack of sleep, and my hands are iodine-stained and I leave a small circle of straw whenever I sit down. It's still below freezing, and the skin on my face has started to crack with the constant cold.

The first birth this morning was alarming, as it's only the second time I've ever seen a breech, or backwards birth. When I feel inside the ewe, instead of a lamb nose and front feet, there are a pair of sharp, pointed hocks and a tiny dangling tail. Luckily it's a simple job to hook my fingers under the tiny back hooves and gently ease out the lamb.

We have a few more difficult births this afternoon so myself (and the kids) are tired, whinging and covered in stringy bits of afterbirth. Steve goes on shift soon, so I'm praying that nothing gives birth before my day ends at 9 p.m.

Sunday, 22nd April

Hatty has finally given birth. She popped out her three lambs overnight, and when I come in at 6 a.m. she's chuckling away and cleaning them very carefully from nose to tail. They're already up on wobbly legs, nosing up into her udder and wiggling their tails when they get a drink.

Hatty's babies have the typical fuzzy-legged look of a Texel/Mule cross lamb, with the slightly squatter bodies from their Beltex dad. All of them have tiny black noses and long black eyelashes. Hatty thinks they're the best thing in the world. She's gone down a bit in size but is still very broad in the beam.

Bertha doesn't seem any closer to giving birth, and is lying in the straw panting away. If she gets any bigger she won't be able to get up.

Monday, 23rd April

One of our younger ewes gave birth to a dead lamb today. This happens. I think the lamb was born prematurely, as it looks tiny and the wool hasn't developed properly along its spine. The ewe is lying in the straw beside the tiny body. She's cleaned it thoroughly but is now completely ignoring her dead baby. I wonder if she's had a knock at feeding time that damaged her lamb and forced her into labour.

I pick up the little corpse and put it in an empty sheep feed bag. The ewe doesn't get up but lies in the straw, staring straight ahead and completely ignoring me. I put the sad little bag outside the shed, ready to be picked up by the knacker man. We're not allowed to dispose of any dead stock ourselves, and dead lambs are no different.

I want to try adopting a pet lamb onto the mother as she has plenty of milk. We have some very young lambs that haven't yet got the hang of drinking from the Titty Machine.

I pick up two of the smallest and move them into a pen. There's lots of different ways to adopt lambs onto a mother, and each farmer swears by their own tricks. I've watched Steve skinning a dead lamb and fitting the fleece onto another lamb. It never seems to work that well for me, and the dead hide often starts to shrink and smell, which I find rather revolting. Some people swear by soaking the adoptee lamb in different substances, to try and disguise the interloper's scent. Some farmers try salty water, the mother's own amniotic fluid or even strong black coffee, before offering the adoptee lamb to the new mother.

I'm not very experienced, and therefore the only successful way I've managed to get a ewe to accept a new baby is to use an 'adopter crate' for a couple of days. The crate is a narrow, rectangular, open-sided box made from strong plastic. You put the ewe into one end of the box and push her head through the end. She can still get up and down and eat and drink as normal, but she can't turn around to hurt or kick the new lambs. The theory is that once the lambs have drunk from her and the milk has passed through their system they will smell 'right', making it more likely that the mother will adopt them.

This is the idea anyhow.

I move the adopter crate into a pen and haul in the new mother. She stands there for a moment then lies down and starts to eat a mouthful of silage. I gently let the pet lambs into the pen and stand back to see what happens.

They blunder about for a bit, so I push them up towards the ewe's udders, trying to encourage them to drink. One gets the idea straight away and dives in there, tail waggling. The other lamb doesn't seem to realise what an udder is for, and noses around hopelessly.

This is when you need reserves of patience. I gently prise open the lamb's jaws and poke its nose under the ewe's udder. I push the lamb onto the ewe's teat and press gently on its nose, trying to stimulate the sucking reflex. It takes a moment, and then the lamb swallows, and its tail flicks from side to side. I take my hand off the nose and it immediately slips off the teat and tries to suck on a mouthful of wool. Sighing, I grab the baby again and have another go. Another couple of sucks and then it slips off for the second time. I'll have to leave them to it and go and feed/water the rest of the flock. Hopefully the second lamb will get the hang of it all before I check back later.

Tuesday, 24th April

The ewe in the crate seems resigned, and isn't struggling when her new lambs try and suck. The second lamb (christened Ratty, as it looks just like a small woolly rat) isn't feeding as well as its new brother, but it must be getting something, as it does tiny lamb poos all over the pen.

Yesterday I watched a story on BBC's *Countryfile* programme about some new technology that will scan a flock and check the sheep's expressions to see if they have any pain or are feeling under the weather. Sheep naturally have a slight smiling expression, but if they're in pain their nose and muzzle tightens and they look pinched and unhappy.

Opinions among the farmers on Twitter are mixed. Some think that shepherds should know their sheep well enough to be able to pick out illness without using technology. I think the 'smiling scanner' is a good idea. Anything that helps improve animal welfare is surely a worthwhile invention. Sheep are very good at hiding illness, as in the wild they would quickly be picked off by any predators if they showed any weakness.

The next time I'm out in the shed, I have a look at my flock's expressions. They all look cheerful, with tiny sheepy smiles.

'Are you happy?' I ask Hatty, who is lying in the straw with her two lambs tucked up next to her warm, fleecy body. She's chewing her cud and her ears are relaxed, out sideways. She looks up at me briefly to check if I've got any illicit sheep nuts, but then turns back to her babies.

Even Crate Ewe looks happy – she's allowing the lambs to

suck without any protest. I let her out of the crate and put her new lambs into the next-door pen. She immediately whickers to them and spins around trying to find out where I've put them. A good result. I stick them back in with her and she sniffs them all over, while they immediately dive in for a drink. Ratty is now sloshing from all the milk he's drunk, and lies down in the straw, his tiny belly bulging to one side. I'll keep an eye on all of them though, as I've seen adoptive mothers kick out and kill their lambs if they suddenly decide they don't want them.

Thursday, 26th April

Not a good day.

One of the ewes scanned as triplets gave birth to three perfect lambs. I move her into a nice airy pen, so she could clean them all off in peace. I iodine their tiny umbilical stumps and they start to stand on their wobbly legs.

Off I went to make a sandwich for lunch, and when I come back I can only see two lambs tucked up next to their mother, rather than three. I push the sheep's rump to make her move and as she stands I see the little crushed corpse flattened under her body. The ewe has lain on one of her lambs and suffocated it underneath her.

I pull it out and sit with the limp body across my knees.

I think there is nothing in terms of lambing that is more frustrating than this.

The ewe is completely unconcerned, and is chewing her cud next to her other new babies.

Maybe she didn't have enough space? I'd moved her into a bigger 'triplet' pen, but maybe she felt it was too small? Maybe she just didn't want to mother three lambs?

I do not believe for one second that the ewe didn't know what she was doing, and I know there will be some reason that I can't yet see to explain why she killed her own baby.

I put the lamb out for the knacker man and walk away to

ring Steve. I'm in tears on the phone, which sounds ridiculous, but it was a needless death of a perfect animal. There's such pressure to produce as many healthy lambs as possible that setbacks can feel like huge disappointments, and it's easy to lose perspective. Steve comes home and sends me off for a break. I think I'm tired and overwrought.

Saturday, 28th April

It's a better day. Finally the weather has improved. I see some watery sunlight, and it feels warm in the yard outside the lambing shed.

I sit down with my back against a pen and just savour the spring sunshine. A starling is busily swooping in and out of the shed entrance. She's built a nest in an old tin can wedged behind the shed pillar and she must have lots of chicks. I can hear them squawking and chattering to each other as she bounds in and out with bugs in her beak.

All the sheep are bedded up and everyone is fed and watered. No ewe is on lambing, and it's wonderful to have half an hour to myself.

I stretch out my feet in my wellies and push them into the warm straw.

In previous years I would keep working for clients through lambing to keep some money coming in. I remember doing a conference call with a client in London while sitting in the lambing shed. It had great 4G mobile connection, and I decided against going back to the house to do the Skype meeting, but instead patched in while perched in the straw.

I think everyone found it distracting, as lambs kept bounding up to the screen and trying to eat my mouse cable. I'd be on Skype telling everyone something sensible about weekly targets when a huge furry face would completely fill their computer screens.

'Sorry about that. It was just Tilly the sheep coming to see if I have any sheep nuts,' I'd say, trying to push the ewe off with one foot while her lambs stood on my keyboard.

It came to a head when I was presenting a short section about a marketing database, and a sheep started nibbling on my 4G box so that I disappeared from the screen mid-flow. I think everyone was grateful when lambing finished and I started working from the spare bedroom again.

Sunday, 29th April

Bertha is finally in full lambing mode. She's scraping at the straw, getting up and lying down and starting to push.

A dark bag of liquid appears at her hind end and bursts, and she turns around to mouth and sniff at the wet straw. She wrinkles her top lip and starts stargazing, stretching out her head and neck to stare at the ceiling. Eventually a foot appears, and then everything shudders to a halt. Bertha is still straining but there's no progress, just a foot sticking out of her vulva.

I kneel behind her and feel inside. Bertha is tired and makes a heartfelt groan while laying her head down into the straw. I can feel the second foot tucked down so I reach through the cervix just as Bertha's contractions hit again.

'Ow! Bloody hell!'

She starts to push and her cervix crushes against my wrist. There's nothing to do but wait it out as the muscular contractions try to expel my hand from her body.

When the pressure and crushing sensation is over, I quickly hook a finger around the tiny hoof and draw the lamb out onto the straw.

She's got two more inside, so I give her a hand with those as well, grabbing the forelegs and guiding the lamb's nose and head through the cervix and out into the fresh air.

Afterwards all three lambs lie wet and shivering on the straw, starting to shake their little heads and make tiny 'meep, meep' noises. Bertha is pleased with her babies and starts to make a chuckling sound while licking them all over.

My hand hurts. I give it a shake, splash iodine over the lambs' umbilical cords and move the little family into a pen.

I need a proper cup of tea. Not one served from a grubby urn. The brewery is open today, so I stagger in, covered in blood and gunk. Heather serves me a cuppa without even wrinkling a nose at my stained waterproofs. I sit at the bar picking bits of straw out of my hair while Lizzie does a wedding 'show round' to a prospective couple. I must look like the local madwoman.

Tuesday, 1st May

My best friends Sarah and Debbie have turned up to give me a hand. Friends often pop in to see me, as they know the tea urn is always on and that I can't go far from the lambing shed.

They bring cake, so we sit on upturned buckets and scoff chocolate gateau while watching the lambs race backwards and forwards in the big pen. The chickens can spot a snack from twenty paces, so I throw them chocolatey crumbs to stop them doing smash-and-grab raids on the confectionery.

I love my friends. They keep me sane and make me laugh. Today I'm feeling really knackered. It's that mid-lambing slump, when you know that half the flock has lambed, all the novelty has worn off, and you still have half to go. Plus you're grey, your skin has gone all lumpy and your hands are cracked and sore.

Sarah and Debs sit patiently with the pet lambs, pushing them up to the Titty Machine and teaching them how to suck.

Wednesday, 2nd May

The Crate Ewe has really taken to Ratty and his brother. They're the cleanest babies in the shed, and she follows them around devotedly. If another ewe wanders over for a sniff, they get immediately butted out of the way. She stamps her front feet ferociously at Sumo the farm cat while lowering her head as a warning. When the little family lies down she positions herself tight up against them and always has one eye open in case of danger.

One little hogg has just squeezed out the smallest lamb I've ever seen. The kids christen him Titchy, and as well as being a bit undersized, he has bandy legs and an undershot chin. His mother thinks he's wonderful and licks him from nose to tail. Titchy is so small that he trips over in the straw, and I keep having to fish him out of the hay rack, as he's too tiny to climb out.

Thursday, 3rd May

Titchy is staying in the pen with his mum a little bit longer than normal. I'm worried that if I turn him out into the nursery paddock I won't be able to see him over the top of the grass.

He's a little fighter though, sucking well from his mum and seeming to grow every day. He gets lots of 'awwwws' from brewery visitors who pop in to see the lambing shed.

Friday, 4th May

Steve's back has 'gone'. At the worst possible time. I find him curled up on the lounge carpet like a prawn.

'I'll be fine! Don't ring the doctor. I haven't got time to go. Just give us a hand up.'

I pull him up and he staggers over to the sofa like a geriatric. Steve won't normally take tablets, but I make him swallow two ibuprofen, two paracetamol and two dusty old codeine tablets I find at the back of the medicine cupboard.

Half an hour later he is extremely cheerful, and hobbles off to check the sheep, singing a song under his breath.

The tablets do the trick for about an hour. In that sixty minutes, a medley of Iron Maiden's best hits is belted out of the sheep shed as he waltzes around cleaning pens. Ben and Lucy think that Daddy is very funny, but it's back to grumpy old bugger when the tablets wear off.

He won't go to the doctor, but I might be able to persuade him to see the guy who looks after horses' backs who comes recommended by a few local farmers.

Saturday, 5th May

Now that the mad rush is over, I invite a few friends to come down and see the lambing, and at one point, have about eight or nine of Ben and Lucy's friends peering at me over the shed fence.

Of course it's Sod's Law that when you have onlookers something unfortunate happens.

I spot a ewe 'hanging' a lamb. She's managed to push out the head, but the baby is caught at the shoulders.

I try to pretend everything is normal, while the labouring ewe careers around the shed, avoiding every attempt to catch her. It looks appalling, as the half-born lamb hangs down half in and half out of the ewe, and the baby becomes more and more purple and swollen the longer it's stuck.

I'm red and sweaty and trying to pretend that chasing a ewe round the shed with a lamb hanging out of her bottom is a normal thing to do.

At 'oohs' and 'ahhs' from the audience I manage to corner the ewe and fling myself at it to bring it down. It's pretty simple just to ease the lamb's head and shoulders out, and in a few moments it's shaking its head and being mothered and cleaned up by the sheep. The swelling and odd colour will wear off over the next few hours.

All this to a running commentary from the watching kids.

'Eurgh that's disgusting! Look at all the blood! And it's doing a poo. Out of its bum. Muuuuuum, it's doing a poo just on the floor!'

'Why have they all got lambs in their tummies? Will they all come out at once?'

'Why is that lamb not coming out? How did it get in there?'

Afterwards I'm washing my hands and one tiny child sidles up to me and mutters, 'Why don't you just hit them with a stick until the babies come out?'

I'm speechless. I stutter a reply about how we '*never* hit our animals, especially when they're having babies'. The little boy shrugs his shoulders and shuffles off. I'm going to have to watch that one ...

Sunday, 6th May

Today I'm wearing my Appalling Lambing Outfit No. 3. It's a huge luminous jacket with a hood that has a sharp point like a triangle. It was left behind by a friend, so I snaffled it. It's for a very tall, very wide person. Which is a bonus as I can wear three coats underneath to keep myself toasty.

I don't usually care what I look like during lambing, but a very nice friend comes to visit and takes some photos of her children on the farm that then appear on Facebook.

In one picture she's taken all you can see is the left side of my face – so one bloodshot eye, a huge jowly chin and my pointy hat. I look like an ancient, demented farmhand.

Monday, 7th May (May Day)

I always feel I'm letting the side down if I don't join in some ancient rural May Day ritual. Are we supposed to roll around in the early morning dew, jump over a bonfire or thwack ourselves with a bunch of daffodils? Will this ensure that our harvest doesn't fail? You never know with archaic rural superstitions.

Instead I crank up our ancient petrol lawnmower and sweat buckets as I drag it around our patch of overgrown back lawn, ignoring Scabby the sheep, who is trying to squeeze her big woolly body through the gaps in the fence to get at the lawn cuttings.

Tuesday, 8th May

This afternoon I'm knackered, and not much is happening, so I find an old picnic blanket under some sacks, wrap myself up in it, and lie down for a snooze on the cleanish straw at the back of the lambing shed. I also discover a stash of Jammie Dodger biscuits hidden under a bucket and have a peaceful time scoffing them before falling asleep.

Wednesday, 9th May

Lambing has slowed down, which is a bit suspicious as there's still quite a few to go, so maybe there will be a huge explosion of lambs in the next few days. Steve's back seems to have recovered, and he's now able to lift the big bags of sheep feed. I let him get on with it and go and sit in the lambing shed.

The latest lamb has something the matter with his front legs. Some of them are born with contracted tendons in their limbs, due to them having been tucked up inside their mothers for so long.

The lamb's lower leg is bent, so he walks on his tiptoes instead. It should sort itself out as he gets stronger and his tendons stretch and lengthen. I've seen some farmers carefully fit wooden sticks as splints. One farmer we know uses wooden spoons strapped to the bent legs with brightly coloured bandages. Each time she goes into their pens all these tiny lambs clatter over with spoon bowls waving above their backs.

Thursday, 10th May

Another busy day of lambing. Tired, fed up and filthy. All pens are full, and as soon as we turf one mother and lamb out into the nursery pen, I have to muck out the dirty straw onto the huge heap outside, and then bed it up again straight away.

It's turned cold yet again, and the bad weather means we can't turn out mothers and lambs, so they're stuffed into every barn or shed corner we can find. Candy the pony has been turfed out of her stable for the benefit of some ewes and lambs, and is shivering at the bottom of her field. I was seriously considering cleaning out the allotment polytunnel, as I reckon I could fit three ewes and their offspring inside.

The cold weather also means that the grass is slow to grow, so we're having to feed the sheep that are outside additional sheep nuts. And that costs more money. This is proving to be a very expensive lambing year, especially as Steve is still out of work. It's all so stressful and I'm tempted to sit in the nearest pen, push the lambs away from the heat lamp and go to sleep in the warmth.

Friday, 11th May

It's cold but dry today which means that Steve has been on his tractor since 6 a.m., trying to catch up with land work. He's ploughing and drilling the next crop of spring barley. He needs to get it into the ground before the rain hits us in the middle of next week.

We're both running on depleted reserves and I'm bloody sick of it. The kids are keeping me going, and my nephew has come up to help. Alex is 15, and a big, strong lad, so he's immediately deputised into mucking out pens. It's so great to have another pair of hands.

Saturday, 12th May

Alex has gone home. We tried to bribe him to stay longer with promises of chocolate cake and bacon sandwiches, but he has to go home to do his GCSE mocks. I tell him that if he gets more experience lambing, he'll be set up for holiday work for life, as farmers are desperate for experienced, sensible night lambers. I think he wants to be an aeronautics engineer instead. Damn.

Sunday, 13th May

Today Steve is ploughing the 'buffer strips' on the farm. At High House these strips are in the eight-acre and fourteen-acre fields, in the wet spots right next to the two streams that run at the bottom. The strips are sown with a special 'pollinator' mix: a blend of colourful wild flowers that will encourage bumble-bees and butterflies and stop any of our pesticides or fertilisers leaching into the water. I pinch a handful of the seed and fling it around my garden to try and perk up the flower beds. Only a small wooden fence separates my garden from the crop fields, and I'm used to having the odd patch of wheat appearing among my geraniums.

Monday, 14th May

A ewe is marked as having twins, and she is starting to push, but there is no birth sac or fluid hanging down behind her. Her cervix is open, but instead of a bag of amniotic fluid, all I can feel is a strange dry mass. I start to pull on a front leg and almost pass out in horror when it comes off in my hand.

The smell is appalling. A mix of rotting flesh and decaying blood. I make an inarticulate sound and turn to retch into the straw.

Disgusting. Poor ewe. The lamb must have died inside her

and I'd inadvertently pulled some of the carcass out. Steve takes over and pulls out (in bits) a mummified lamb.

The smell is everywhere, and I wrap the corpse and put it out for the knacker man. Incredibly, Steve can feel a healthy, if small, single behind the dead lamb, so he successfully reaches in and pulls it out and puts mother and baby into a clean pen.

We've given her a massive dose of antibiotics, painkillers and anti-inflammatories. I've washed myself twice in the shower and I still can't get rid of the smell.

Tuesday, 15th May

I still smell revolting. Nothing is shifting it. It's a mix of dead lamb, spoilt silage and amniotic fluid. Mum offers me a lamb hotpot for tea, and I can't face it, as it smells the same as inside the lambing shed. I eat cheese on toast instead.

Wednesday, 16th May

Around lunchtime a pair of lambs are born, but because I didn't get to them fast enough, one of our Suffolk ewes has 'pinched' one of the newborns.

The Suffolk is pregnant herself, but is so crazed on prebirth hormones that if she sees a lamb born she'll take it off the real mother and convince herself she's given birth. Except she's still full of her own unborn babies and hasn't got any milk.

By the time I get to the shed after lunch, one lamb is being suckled by the proper mother, and the other lamb is standing under the very smug Suffolk 'pincher' ewe, headbutting her udder and blaring for milk.

Of course, the real mother refuses to accept back her second lamb now that someone else has licked it.

I stuff the real mother in the adopter crate and manage to

eventually drag the pincher ewe away from both lambs. Then I run to pick up the lambs and plonk them back in with their real mother in the pen.

A day or two in the crate will hopefully let the real mother re-bond with her babies.

Pincher ewe is now trying to squeeze her entire fat body through the gaps in the pen gate, bleating in desperation to reach what she thinks is her lamb.

Everyone is confused, fed up and bored with the pincher ewe's dramatics.

Eventually I get so annoyed at the noise that Steve helps me move the real family into another shed, and eventually the pincher ewe settles down for her tea.

Hopefully she'll lamb herself in the next few days, and the problem will disappear.

Thursday, 17th May

I'm tired today. I'd forgotten to take my watch off before my shift in the lambing shed last night. It disappeared halfway through a lambing, and for a moment I thought I'd actually lost it inside a ewe. Fortunately I found it this morning when I was mucking out a pen.

Saturday, 19th May

Titchy, who is our most miniature lamb so far, is managing very well. He stands on tiptoes to reach his mother's milk, but he's drinking lots and spending much of his time lying on his mother's back. It must be the softest and warmest place in the whole pen.

I've spent the afternoon trying to stop the Suffolk pincher ewe from nicking everyone else's babies. I think she's gone completely mental. She's started licking lambs as soon as they emerge from another ewe's bottom, which really pisses every-one off.

After wrestling her to the ground a few times I give up and shut her in her own pen. She's now yelling over the side and scuffling in the straw trying to find what I can only imagine is her own imaginary lamb. Completely barmy.

Sunday, 20th May

The Suffolk pincher ewe has finally lambed, and therefore stopped trying to kidnap everyone else's babies. She has twins that match her exactly, with black heads, huge satellite ears and splotchy black-and-white legs. She's a brilliant mother, as she should be with so much bloody practice.

Monday, 21st May

We have finished lambing!

Everyone (including the twelve hoggs) have finally given birth. Spotty Nose was the last ewe to go into labour, giving birth a full three days after everyone else. She'd been marked as expecting twins, but instead managed to cough out one skinny single lamb. Mother and baby are both sitting under the heat lamp, tucked up together in the straw.

Spotty Nose is our oldest sheep in the flock at six years old. She's not the most attractive lady, and has a misshapen nose and lots of black freckles and spots. She must have been mated by the tup on the third cycle, rather than the first or second cycles like the rest of the flock, hence being the last to lamb. Steve reckons that the tup went around everyone else first, and then got to Spotty ...

Tuesday, 22nd May

Now lambing has finished we have a tally of:

- Twelve pet lambs. Most of these are now great hulking animals, and are eating lamb feed and don't need any milk. The slightly smaller or dimmer lambs are still on the Titty Machine. These include Fred, Fuzzy and two tiny spotty brothers whose mother didn't have enough milk to feed her pair of lambs. The Spotty Brothers are covered in a pretty pattern of black and brown splotches. Fortunately, they get the hang of the Titty Machine without too much coaching, and spend their time curled up together with two fat, full tummies.

- Around 290 bouncing healthy lambs, although I haven't counted them yet. These include Titchy – the minuscule lamb that was born to our smallest hogg. They're currently outside with their proud mothers, and don't seem to be taller than a blade of grass.

Everyone lets out a huge sigh of relief, and I lie on the carpet with a cup of tea for a while, mentally congratulating myself for getting through the biggest lambing I've ever done without a) divorcing Steve b) accidentally killing any human or animal and c) going mad through sleep deprivation.

Thursday, 24th May

There's still plenty of work to be doing, even though lambing has finished.

The Titty Machine needs cleaning every day, which is a job I absolutely hate, and I'm still checking ewes and lambs outside and inside, castrating lambs and feeding everything twice a day.

But at least I don't need to muck out lambing pens any more.

I think I've lost about half a stone, and my hands are red and cracked, no matter how much E45 I slather over them.

Steve's back is still sore, and we haven't sat down for a proper dinner in about six weeks.

Steve is now in a race to get all the spring barley into the ground while the weather is dry. He loves sitting on his tractor, and if everything is going well, he falls into a zen-like state. I find trundling up and down a field for hour after hour deeply tedious, so I leave him to it.

Friday, 25th May

Our oilseed rape is in full flower. The buds are a bright acid yellow, which isn't to everyone's taste, as it's such a contrast to the pale green of the grass fields, but I rather like the shimmering splashes of colour. We have a rape field right next to the house and the honey fragrance has started drifting through our open windows. I go for a walk around the field edge to look at the crop. Close up, the rape loses the intense sweet scent, and instead has a heavier, astringent smell. The field is abuzz with activity, honeybees hovering in among the flower heads and small cabbage white butterflies fluttering from plant to plant.

The ground around the edge of the field is tricky to walk on, but I stumble over the plough marks and see many tiny vole and mice holes in the cracked soil. No wonder the buzzards like to nest in the corner of this field. It must be teeming with their favourite sort of food. Sumo the farm cat also likes to hunt in among the crop stalks, and I often see him, intent on some unseen prey, black-and-white tail swishing from side to side.

When I walk back to the house my trouser legs and wellies are covered in yellow pollen. The entire family (apart from me) suffers badly from hay fever, so I do the kind thing and wash off my boots at the outside tap before I go inside. I can't even have flowers in the house, as it sets everyone off sneezing and complaining.

Saturday, 26th May

It's celebration day as we move all the pet lambs outside and turn off the Titty Machine.

It's warm and sunny, and we attach the trailer to the quad bike, and gently move them all into the sheltered paddock in front of the house.

All twelve of the pet lambs are now sitting squeezed up against the outside shed wall with big eyes, staring at the field and sky. They've only known the lambing shed so far in their small lives, and being let out into a paddock has blown their tiny minds.

One of them occasionally gets up enough courage to go and nibble on the grass.* The Spotty Brothers sit tight together, noses pointed to the sky in typical 'lamb comfy pose'. Fuzzy is close beside them, tucked up against the fence with a bewildered expression. He smells a bit better now. He keeps shutting his eyes, maybe to block out the view of the paddock. Hopefully they'll get brave enough to do a bit of grazing before long.

Sunday, 27th May

Last night a ewe died in the field, due to rolling on her back and suffocating.

When we found her she'd been dead for a number of hours.

---

* Usually lambs stay with their mothers for around twelve weeks, until they are fully grown. As well as drinking her milk they start to eat grass from day one. Our pet lambs don't have a mother, so at six weeks, we shut down the Titty Machine in the shed and they drink water and eat lamb feed, hay and eventually grass, when they're turned out into the paddock.

Ben saw her and reported delightedly, 'She's dead! She's dead! Daaaaaad, she's even gone all bloaty!'

He wanted to poke her with a stick. This is his 'go-to' behaviour when seeing anything slightly disgusting.

The dead ewe left behind a pair of bonny lambs.

Heartbreakingly, once we take her out of the field in the loader trailer, the lambs sit tight on the ground where she'd died. I've just been down to check on them, and they're still in the same spot.

Monday, 28th May

The pair of orphaned lambs seem to be following around another ewe. She's letting them sit close by to her, and although she won't let them drink her milk, they look a bit happier. I hope they're mature enough to cope with just a grass diet. We can't give them any supplements: I don't want to chase them to catch them to bring them inside; it would mean we'd have to feed every other lamb in the same field.

Tuesday, 29th May

All our lambs are outside enjoying the sunshine, basking in the warmth. The orphaned twins have survived the night, and are tucked up close to their Aunty Ewe, with their noses pointed towards the sky. The pet lambs have their heads down in the grass enjoying their breakfast.

I spot someone walking through our front field. They have a dog on a lead, and another running loose.

We do have signs. Polite ones that I bought especially, asking dog owners to 'Please keep your dogs on a lead as we are in lambing season'.

The dog walker reaches the gate and I see that he has two black Labradors.

'Excuse me, we're just at the end of lambing and I was wondering whether you could keep your dogs on a lead?' I ask, mentally thankful that Steve isn't here, who would have used much shorter and nastier words.

The dog walker turns and looks me up and down. He's dressed for the warm weather, while I must look a right state, in stained farm overalls and filthy boots.

'My dogs always come to call,' he snaps sharply, pulling one dog to him by the collar, and clicking his fingers at the other.

'Well, that's great, but our sheep don't know that,' I say sweetly. 'And they panic if they see a loose dog, and it makes them leave their lambs.'

The dog walker doesn't say anything, but just marches on up the drive, his dogs following him at a short distance.

'Thanks for that *you complete and utter bellend!*' I shout as he walks past the brewery.

He shrugs his shoulders and continues up the path.

I wonder if I'll become the subject of an irate Facebook post about a 'very rude farmer who confronted me on a public footpath'. Most people counsel farmers not to lose their tempers and remain polite, but I'm tired and filthy and I keep seeing police posts on Twitter and Facebook with gory images of dog-worried sheep. Even the calmest family pet dog can go postal and start chasing stock, and I'm buggered if that's going to happen to my ewes and lambs.

I'm shaking. I'm not very good at confrontation. Having a public footpath through our farm is both a blessing and a curse. A blessing as it brings business to the brewery and the wedding venue. A curse as some leave gates open, break stiles and drop litter. Or allow their pet dogs to roam free, ignoring my extremely polite notices.

Thursday, 31st May

This evening I spend some time leaning out over my window-sill listening to the curlews and watching our sheep grazing in the back field. A few lambs are sitting with their heads tipped back, enjoying the spring sunshine.

I spot a hare crouched in the long grass beside the field gate. All I can see is a pair of twitching brown ears as he hunkers in a scrape in the ground.

After a while the hare tentatively moves into the open and comes towards the gate, lolloping through the wooden struts and onto the gravel drive.

He's beautiful. I've never seen one so close up before. His long, slim ears are shell pink inside, and shade to a charcoal black at the very tips. His oddly human eyes are deep amber and outlined in black, while his thick coat is speckled with different tones of beige and brown.

I hold my breath, trying to keep as still as possible, as the hare hops across the grey stones. I can hear the delicate 'chink, chink' as the gravel moves beneath his paws, and I see the occasional flash of a pure white belly and chest.

Just then, one of the children makes a noise in the house, and the hare freezes, then bolts across the gravel and back into the field. He doesn't run in a straight line, but 'jinks' from side to side, as if he's running away from an imaginary pursuer. Maybe he was looking for a mate?

We regularly see hares 'boxing' in our front and back fields, and it's common to see them run across the road in front of our cars, but I've never seen one come so close to the house.

Friday, 1st June

Even though lambing has finished and all the dreadful stress has passed, I suddenly have an attack of anxiety. Everything assumes the proportions of a disaster. The future looks bleak. Steve will never get another job. We're all doomed.

I've no idea why my mind does this to me. It waits until I lower my guard and then all the panic and neuroses come flooding back.

I spend the day doing unhelpful things. Eating sugar. Sleeping too much. Stressing about how overweight and unfit I am. Not having a shower. Obsessively checking social media, and scrolling through other people's Instagram posts to peer at their perfect and worry-free lives.

After a couple of days of this I reach rock bottom, and force myself out for a walk. I trundle down to the wood and spend thirty minutes marching through the brambles, pushing myself to examine the trees and listen for birdsong, and to smell the daffodils. My mind is still chattering at high speed, but being outside gives me a glimpse of perspective.

Of course, life still goes on. I can't stop feeding the children, looking after the sheep or doing the housework just because I'm anxious. To an outsider, I look OK. Probably a bit jittery and tired, but still coping and being an adult, doing grown-up things.

When I'm this anxious I escape by reading lots of books. I read voraciously, two to three books a week, gulping down novels and nonfiction and anything I can get my hands on. Losing myself in someone else's world helps me to anaesthetise my brain and stop the chattering of real life. But it's only a temporary relief; when you turn the last page or switch off your bedside light, the insidious anxieties creep back, often stronger than before.

Steve has been looking for jobs and has been on a couple more interviews. He'd have no problem finding a great full-time job as a site manager, but part-time work is few and far between. Our finances are stretched to the limit. My car has also just failed its MOT and according to the garage, the repair bill will be well over £400. I can't afford to have it fixed, so with a heavy heart I tell the DVLA that I've taken it off the road. I'll just have to beg lifts until we can manage to scrape together the money to run it again.

Saturday, 2nd June

My anxiety is so bad that I'm finding it hard to get out of bed let alone the house. I carry around a feeling of impending disaster, and tiny setbacks make me burst into tears.

Eventually Steve persuades me to see the doctor, and after an appointment with the local surgery I come back out with some new antidepressants and a recommendation for mindfulness and more exercise.

The feed bill from lambing plops onto the mat and it's much bigger than normal, due to the extra sheep feed we've been giving the sheep to keep them going through the horrible weather.

I'm not sure what to do but begin taking the tablets and force myself outside for a walk every day.

Sunday, 3rd June

We're not the only ones feeling the pinch. Poor grass growth, big heating bills and no money coming in from crops or sheep has pushed many local farmers to the brink. At least the kids have free school meals and Granny and Grandad to help them out with uniform and shoes.

In desperation we ring the Royal Agricultural Benevolent Institution (RABI).* It's embarrassing to have to ask for help, but we're not sure what else to do. The nice man at the end of the RABI helpline asks lots of questions. He tells us that mental health in farmers is at an all-time low, and he has personally spoken to a lot of people in agriculture who are finding it hard to make ends meet. This makes us feel better, knowing that we're not the only ones. He promises to come out soon and have a chat to see what they can do.

This isn't something I would normally tell people, but

---

* RABI is a welfare charity that exists to help farming people in crisis and financial difficulties.

once I start sheepishly mentioning on Twitter that we're going through a rough patch, the kind messages start flooding through. Other farmers tell me how close to the breaking point they've been, and lots of people send such lovely emails that I'm in tears.*

Tuesday, 5th June

The grass is really starting to 'come away'. We've had a couple of days of rain, which has put a stop to the ploughing and drill-ing, but today I can see that the sudden moisture has provided a flush of new grassy growth.

The pet lambs in the paddock are beginning to grow into chunky little individuals, and the ewes in the big fields are looking less scrawny. Pregnancy and producing milk make a sheep lose condition, but good grass is the best medicine, and really makes a difference to their weight.

All the sheep need treating with fly repellent and dosing with wormer, and the ewes need dagging, to remove any dried-on 'poo danglers' from their rear ends. The repellent and dagging will hopefully stop any fly strike, and will make things easier for our shearer when he comes in the next few weeks.

Fly strike is very unpleasant. It can happen overnight, and the first sign is an itchy sheep who can't stop scratching against fence posts or stone walls. In warm weather flies start to breed, and they will lay their eggs in the warm, smelly fleece around a sheep's bottom. The maggots hatch and burrow into the skin, making the sheep scratch, breaking the skin even further. This is the reason that farmers dock the tails of lambs, and one good reason to shear sheep's fleece. If tails and fleece are left long

---

* We debated long and hard whether to include this entry in the diary. It's embarrassing that people will know how desperately we needed help. But then I thought – bugger it. In my opinion it's important not to sugarcoat how tough it is to make money in farming. It's also important to show anyone else who is struggling that it is OK to ask for help.

they get covered in faeces, and are a perfect target for egg-laying bluebottles. If it's left too long a sheep will lose weight, and can even die from blood loss and septicaemia.

We treat the sheep before and during the fly season, and keep a close eye on the flock during the warm weather.

Spraying, dagging and worming is a long job with a big flock, so the earlier we start the sooner we'll finish. Also, we don't want to be moving sheep during the hottest part of the day as it makes them too hot and stresses their systems.

Mavis is sent round the back field to gather the flock together. It's a tricky job, as the lambs get confused and lose their mothers, and then the ewes start bleating for them, and refuse to move towards the gate. Eventually, after a bit of arm flapping and shouting, they're all gathered into the farmyard and out of the sun.

Even after that short run the ewes are panting heavily. Their fleece makes them overheat, and the sooner we can get them sheared the better. Except our shearer, John, is in great demand, with plenty of other farmers needing his services, so we will have to be patient and wait in the queue.

Then the work begins. I drive a few lambs and ewes at a time into the barn, and Steve catches each lamb and gives them a dose of the worm drench. I don't suppose it tastes very nice, and there's a real skill to grabbing each lamb, pushing the syringe properly into their mouths to make sure it goes down their throat. Each ewe needs to be caught and her head put into a circle of twine, so that Steve can use the electric clippers to cut off the soiled fleece at her rear and spray her along the back and sides with repellent.

It's a sweaty, dirty, backbreaking job. Sometimes a ewe's backside is absolutely plastered in green poo as a result of eating fresh grass. Steve's hands are slippery and at the end of the morning he's aching and splattered from head to toe with dung and worm drench.

I take over while he eats a sandwich and has a cup of tea. The lambs wriggle in panic and lunge away when I try to hold them still to squirt in the wormer. These aren't cute little babies any more, but thick-set, muscled animals, with some real strength behind their snub noses and cute faces. I'm sweating while I struggle to catch each one and hold them still. While I'm bent over trying to give a lamb a dose of wormer, another one jumps past at head height, giving me a right crack against the side of my skull. I sit down suddenly and drop the wormer, spilling the white medicine all over the floor.

It's not serious, but I have a rather good bruise on my right cheekbone. Steve sends me out for a bit of fresh air and a cup of tea while he carries on with the rest.

I sit on the cobbles with my head in my hands. Sometimes I think I'm not much use. I know it's not all to do with strength, but rather the knack of knowing how to hold and immobilise a lamb, so that it can't pull away from your grasp. Even so, sometimes I wonder if Steve would have been better with a great strong farmer's daughter as a wife, rather than a weedy shortarse …

Wednesday, 6th June

All the ewes and lambs have been dagged and dosed, and now they're out in the fields, head down, munching through the grass. We'll need to dose the lambs again in another few weeks, but for the meantime the flock is peaceful and enjoying the sunshine.

The mark on my cheekbone has developed into a full-blown black eye. The kids inspect it solemnly and Ben asks me to tell him again how the lamb bashed into me. I must remember to watch the children around the animals. Not so long ago, Lucy was helping us move some sheep into the ring at the Mart, and a ewe ran into her from behind, knocking her off her feet and giving her some nasty bruises up her legs and spine. They can be rough, dangerous animals, especially with children handlers.

Thursday, 7th June

Today we're nervous, as someone from RABI is visiting the house.

Once the kids are at school we gather our bank statements and bills together and wait for the appointment, biting our nails and drinking endless cups of tea. When the RABI rep arrives he's very nice and sympathetic, and carefully goes through our accounts to give an outline on how he can help.

RABI doesn't assist with business finances, but will help us with our household bills, such as heating oil and clothes for the kids. He's lovely and discreet and I babble on to him about how hard lambing has been, and farming in general. He nods and says all the right things, but when he leaves I walk into the lounge to find Steve sitting on the sofa quietly, his head in his hands.

I know what he's thinking. Steve prides himself on being able to provide for his family, and the knowledge that he's not making enough money to even pay for our weekly food bill hits him very hard.

We've done everything we can – stretched finances, invested wisely, cut back all our expenses – but we just can't make ends meet. We sit in silence next to each other, each one with our own thoughts, until it's time to go back out to check the sheep.

Friday, 8th June

Good news at last. Steve has managed to find a part-time job. It's not a site manager's job but rather working at the local agricultural merchant's in Hexham. He'll work three days a week and then fit the farm work into the remaining four days. At least now he'll have a regular monthly income; we celebrate with fish and chips from the local chippie.

We tell an elderly farming friend. 'But why do you *need*

another job?' he says. 'Won't you be embarrassed to be working behind the counter, to be serving other farmers?'

I don't think they understand how desperate our finances have become.

Once the kids are in bed we sit down and work out a budget. There will be no money for luxuries and new clothes, but the relief of having a regular income makes us leap up and dance around the kitchen. The children don't understand what's going on, but they lean over the bannister and laugh at us waltzing around the dining room table. Things might be looking up.

Monday, 11th June

RABI have sent us a letter. It reads: 'I am sorry to learn of the difficulties you have been facing and wish to help where we can. We are unable to help with business bills but can give a grant towards domestic costs.'

They have enclosed a small cheque to 'assist with the purchase of heating oil and help with further household bills'.

We are very grateful. It feels mortifying asking for help, and I don't suppose anyone knows how much we struggle with day-to-day costs. From the outside, we look on an even keel. We run a car, the kids are in clean clothes and have enough to eat, but it's all a very thin veneer. I never thought we'd have to turn to charity to survive, and I feel a curdling mixture of gratitude and shame as I look at the cheque.

Steve and I write a heartfelt thank you letter and carefully put the cheque in a safe place to pay into the bank at the earliest opportunity.

Wednesday, 13th June

I tripped over a straw bale last night and landed heavily on my outstretched arm. My elbow made a twanging noise and started to inflate. Ow. Steve took me off to Hexham Urgent Care Centre, where, according to the media, we should be prepared to battle through the pale hordes of flu-ridden undead, gnawing on toilet rolls and drinking the hand sanitiser. But I was seen in thirty minutes. And the hospital was so clean.

There was an elderly man in the waiting room. He'd hurt his finger in his wheelie bin, and we had a good chat, bonding over a chocolate finger from the vending machine.

Dr Abdul was the main doctor on duty. He has treated my family through two gall bladder attacks, a suspected miscarriage, Steve crushing his hand in a combine header, Ben chopping off his fingertip in a door, Lucy falling off the fat pony and toppling off a hay bale, Steve slicing his scalp open with the PTO tractor guard and accidentally ramming the foot-trimming shears into his hand.

Dr Abdul is from Turkey, speaks heavily accented English and is a Hexham institution. Every time I go to Hexham Hospital he's there, but looking a bit older and wearier each time. We need to give him a medal for services to the sick and injured of Tynedale.

There are rumours that the Hexham Urgent Care unit will be shutting, but it's an essential part of our household, and I bet an essential part of other local farming businesses.

My elbow was inspected and declared fit, and I was shown the door, all in around an hour. The nurses were cheerful, professional and efficient, and were interested to know how lambing had gone.

Friday, 15th June

I'm back in Hexham Urgent Care again. It's completely my own fault. Today we were inoculating the last few pairs of lambs with a dose of Scabivax Forte. We use Scabivax to prevent orf disease,* and I was standing in the lambing pens, bending over each pair of lambs, carefully scratching them on the inner part of their thighs with a loaded scarification needle.

A ewe took umbrage to me lifting up one of her lambs and thumped me with her head, just as I was straightening up, needle in hand.

My arm flew up and inadvertently scratched my ear. I then gave it a right good rub with a dirty finger and carried on. Over the next few days my ear started to throb and go yellow. Today it's leaking pus and is pulsing ominously.

I troop back to see Dr Abdul.

'You injected yourself with what?' he asks, bewildered, as his English isn't that great, and my Turkish is non-existent.

'Orf vaccine. To stop orf in sheep. When their skin goes all blistered and bumpy,' I reply.

He inspects my ear then tuts, shakes his head and goes off to ring the NHS to ask when I last had a tetanus injection. I sit on the hospital bed, swinging my wellies backwards and forwards.

Apparently, I haven't had a tetanus injection since 1992. I get a lecture about keeping up to date with inoculations, then a stinging jab in the arm plus a bag full of antibiotics and some squeezy stuff to put in my ear.

Dr Abdul tells me to see my GP if I suddenly break out in huge orf pustules, and escorts me out of the hospital reception. I'm fairly sure the receptionist rolls her eyes as I shamble away.

I think I can hear cheering as I push open the doors and douse myself in antibacterial hand wash.

---

* 'Orf' is another name for 'contagious pustular dermatitis', which leads to painful spots and rashes. It can also affect humans. We had a friend who caught orf from a sheep and then scratched his bum, and he developed nasty boils all over his bottom.

Saturday, 16th June

We race round the sheep to finish early for the evening, as to-night is the annual village Talent Show. Set up by the vicar, it's a popular get-together, where anyone can get up and do a 'turn' to show off their personal talents.

My friends and I have often discussed whether we could cobble together an act, but lack of confidence means that we always bottle out.

'Why don't you do a reading of some of your blog?' Steve suggests.

I'm too embarrassed. People would recognise me, and there's nothing like sticking your head above the parapet for attracting criticism. Plus I'm still feeling too anxious to do anything to draw attention to myself.

Instead, we all gather in the village hall and sit at long tables. The vicar does a brilliant intro, and we all sing a song. The hall is packed full and we're offered sausages in a bun, a selection of home-made puddings and warm glasses of wine.

The local art group is displaying their paintings, while the crafting group has a table covered in intricate embroidery and crochet. There's a raffle with the usual prizes of a bottle of sherry, a chintzy toiletry set and a big box of chocolates. The entertainment starts. Lucy and Ben sit at the front of the audience, mouths open and eyes wide at the acts on stage. We hardly ever go to the theatre or the cinema, so this is a real treat for them.

First one is a chap on the bagpipes. We all clap along with enthusiasm. Following that is a comedy skit, a young lassie singing songs, some tiny kids playing the violin and the key-boards and eventually, my favourite bit of the show – the Armstrong brothers.

Willie and Ted, 85 and 92, sit on chairs and have a bit of a blether about the old farming community, telling stories from the 1940s and 1950s.

Their accents are proper Northumbrian, with a rolling

'r' sound, and they still use words like 'cowp ow'er the dyke' (fall over the dyke) and 'muckle' (big). Willie tells a story about a farmhand who 'used to get a bit angry like when we were bairns. We nivver crossed him, but he'd get into a fettle with summat, and when he got these two red dots on his cheeks ye had to run fast into the hoose before he belted you one.'

Fabulous.

It's a great night. Everyone joins in the singing, laughs uproariously at the sketches and has a fair bit to drink. The kids are bright-eyed with excitement, and don't want to go home.

When we get back sometime after 10 p.m., we all go around the sheep sheds together, repeating bits of comedy skits to each other and giggling like loons.

Monday, 18th June

Steve has brought my car into the shed and is busily trying to fix it himself.

He's ordered all the parts and bought a book off the Internet to show him what to do. I take him a cup of tea and find him underneath the car, clanking away with a screwdriver and an oily rag.

He pushes himself upright and takes the steaming cuppa out of my hands.

'I can't afford to pay for someone to fix it,' he says sheepishly, 'but if I have a go, I might be able to do it myself.'

He takes a swig of his tea then crouches down again to get back under the car.

My eyes get all misty. Steve might not be the most romantic bloke in the world, but he'll work his socks off to keep our little family going.

SUMMER

Summer is all about watching the weather and hoping it will allow us to make great hay and harvest a good crop of oilseed rape, wheat and barley. We're also checking our flock for any heat-related health problems, shearing our sheep for the hot weather, weaning the lambs and getting them ready for sale at the Mart.

Wednesday, 20th June

The weather forecasters are predicting a heatwave for the next few weeks.

We've not had any significant rain for over ten days, and the Sparrow's Letch is now down to a miserly trickle, the water choked by great swathes of dark-green bittercress weeds.

The terrible winter and spring has checked the growth of our grass anyway, and I dread to think what a prolonged drought might do.

Thursday, 21st June

Steve and I are gearing up to do a nekkid summer solstice dance around the hay bales, to ensure a good harvest (with strategically placed chickens for warmth and modesty). Although I'll also have to tie my bosoms in a bow to keep them out of the dust.

I've been watching the crowds gather around Stonehenge on TV, which got me thinking about old farming superstitions. (I was going to get up to take a picture of the solstice dawn over our wheat fields, but I'm crappy at getting up in the morning, and it was cold.)

Some of the farmers I know are still superstitious – but it's understandable when your harvest and livelihood depends on the whims of the weather gods. It's important to keep them happy.

We still know people who can make beautiful corn dollies from the last sheaf of cut wheat or barley. Originally, they were kept safe until the new year, when they were ploughed into the first furrow – a sacrifice to keep the harvest god happy. The Church Christianised the whole idea, so now corn dollies are displayed in the local church for Harvest Festival.

Whalton, a village twelve miles away, used to make a 'kern babby' (corn baby) from the last sheaf of the harvest, dress it up and display it in the church until the Harvest Festival. I've seen an old picture of a kern babby with children dancing around it – the babby is set up like a maypole and looks well over seven foot tall, wrapped in a long white dress with an explosion of wheat and flowers for a head. It's a bit unnerving, to be honest.

Old iron is nailed up over doorways to keep away bad luck, usually in the form of a horseshoe. (We have some huge shoes that must have been used by heavy horses pre-war.)

The wooden beams in our old buildings have the occasional scratched 'VV' mark. Steve thinks they're just carpenter's marks, but they do look very like the 'witch prevention' marks I've seen. VV refers to 'Virgo Virginum', or the Virgin Mary. It's often scratched in plaster or woodwork over doorways and windows to prevent evil spirits entering a building.

When the farm buildings were converted into the brewery in 2003, our builders pressed copper coins into the wet plaster near the roof to ensure good luck. Old traditions die hard around these parts.

Friday, 22nd June

The summer is underway, and every weekend the brewery hosts one or two wedding ceremonies. The kids and I like to hang over the stone wall and watch the bride arriving. Today is a particularly upmarket event. The guests gather outside our kitchen window and we stare at them in their beautiful dresses, morning coats and big hats.

We're becoming experts in wedding dress fashions. For a long time, it was all about strapless corsets, but now, lace is back in vogue, and almost every bride has a beautiful lacey bodice to their dress, with long, silky skirts.

This bride is wearing a long-sleeved lace dress with a circlet of pink flowers in her hair, and her many bridesmaids are dressed in a bright, retro fuschia and orange flowered fabric.

'My nanna used to have that as her bedroom curtains,' says Steve, pointing at the flowery dresses of the bridesmaids, who are now lining up outside our window. I shut the window in case someone overhears. I must admit they do have a certain 1970s sort of vibe, and I'm pretty sure I can remember a childhood bedspread in the same material.

The guests follow the registrar for the ceremony in the hay barn, which has been decorated with big bunches of bright pink flowers, fairy lights and huge cream paper globes. We lurk in our front garden listening to the tinkle of polite conversation and the pop and fizz of champagne bottles being opened and poured.

We must be very quiet when the ceremony is taking place, as it's held in the open-sided hay barn, so we all tiptoe around in the background, feeding sheep and bedding up the pet lambs. Steve tries hard not to crank up any tractors or quad bikes when the marriage ceremony is happening, although during a stressful harvest he has been known to just keep on driving his tractor and trailer right through a wedding reception.

Then the party starts, and we can hear the strains of music and hum of conversation drifting from the brewery buildings. During the summer many brides hire a marquee, and then you can really hear the live bands or the ceilidh dance caller. Some of them are great, some are not so good, but in any case, I'm now word perfect to 'Come on Eileen' and most of the Coldplay back catalogue.

After I put the kids to bed I go out for my late-night check and there's a chap staggering around the farm machines in the barn. He's absolutely hammered.

'I like your tractorshh,' he slurs happily at me. I nod and smile and try to steer him out of the barn. He must have ignored all the warning signs.

'Can I have a go on it?' he asks. 'Go on. The big one. Over there. That one.' He points unsteadily to our black Valtra tractor that's parked up in the corner, while his body takes on a sideways lean.

'Errrr, no?' I quaver, trying to shoo him out of the shed without getting a face full of hot, beery breath.

'Ahhhhhhhhhh, gan on. Gan onnnnn,' he wheedles. 'I'll be ever so careful. I promishhh.'

I manage to get him to the barn door and then tell him that he needs to go back to the party, in case he falls over.

He suddenly hauls himself up to his full height and shouts 'I can go where i likes!'

Fair enough. I'm not about to start fighting with a big hairy drunk. I back away and, uncharitably hoping he manages to impale himself on the tractor bale spikes, go and find who he belongs to. Eventually his embarrassed wife manages to coax him out of the barn and back into the party.

She's ever so apologetic. Warning signs and door locks don't seem to work against the average determined drunk wedding guest. I hope he has a satanic hangover the next morning.

Saturday, 23rd June

On the morning after the wedding Lucy and Ben gallop out to the hay barn to pick over the party detritus.

I discover Lucy has claimed some party hats and has found a stash of unopened confetti under one of the tables. I've stopped trying to prevent them from scavenging for things after a wedding. The kids see it as their haul, and the brewery staff don't seem to mind. If they're not peeling up sweets from the cobbles or chugging half-drunk glasses of champagne, I'm OK.

However, I spot Ben firkling in one corner of the farmyard, turning his body away from me so I can't see what he's doing.

'What is it?!' There's no response. 'Come on, hand it over!' I demand, holding my hand out flat.

Sheepishly he turns around and I see that he's eating a discarded packet of Doritos that have lain half opened next to the sheep feed all night.

'Ben, that's disgusting!' I snap. 'God knows where they've been.'

It's not like I don't feed the kids. They get three solid meals a day plus ad-lib snacks. And sweets too. It must be the thrill of finding 'treasure' left behind by party guests.

When wedding parties lay on children's entertainment for their younger guests I have great difficulty in persuading Lucy and Ben that they can't join in and watch the magician or have a go on someone else's bouncy castle.

'Just because you live here doesn't mean that you're automatically invited to someone else's wedding,' I instruct, as they press their noses inconsolably against the lounge window, watching kids in party dresses bouncing up and down and hooting with excitement.

I did find Ben in his pyjamas having a sneaky go on a pirate bouncy castle last year. At six years old the temptation had become too much, and he'd got up, undone the front door and rushed out in his onesie and wellies to join the other kids. I heard him over the hedge and marched out to haul him back into the house. He had such a telling off that I'm hoping he won't dare to do it again. The children also don't realise, in their innocence, that as we don't know the guests, I don't want them wandering up to them and chatting, as they do with other farm visitors.

Lucy now watches guests like a hawk out of her upstairs bedroom window.

'They're all drunk again Mum!' she calls down to me. 'Good for them!' I shout up the stairs. I'm glad people have a good time on the farm. And if everyone gets a bit merry, I'm not going to judge. When I go to a wedding I like to get my money's worth, and have a right good go at the buffet and bar, until I get too over-excited and Steve has to drag me home.

Sunday, 24th June

I'm balancing precariously on the roof of Candy's stable, legs on either side of the top ridge, holding a brush and long rake in both hands.

'I'm not enjoying this very much,' I say. Steve is sitting on the bottom rung of the ladder. He looks up and shades his eyes against the sun.

'Don't move around a lot. I don't want you to go through the roof. Fixing it will cost a fortune.'

We're clearing out the gutters, as otherwise they clog with leaves and rainwater pours over the top and drips onto Candy and our sheep below. One year we were too busy to dig them out and we ended up with a torrent of water that soaked the straw and ruined a few bales of hay.

I'm on the roof as I'm lighter than Steve, although I'm not convinced that this is the real reason. He doesn't much like heights and prefers to shout up commands while I totter unsteadily up the ladder and balance on the tiles.

There's an actual tree sapling growing in one of the gutters round the back of the building, and I inch down to hoick it out, wobbling dangerously when I throw it over the side. It misses Steve by inches. I start to poke out the clumps of decaying twigs and leaves and haul up the thick carpet of moss that covers the gutters. I throw it into the stable yard, taking great delight in aiming at Steve's head. He ducks under the shelter of the stable door until I've finished, and then hands up the handle of the power washer. This is much more fun. I scour out the pipes and gutters, managing to accidentally soak Steve a few times as well. The job is finally done, and I shuffle my way back to the ladder and creep down the rungs, inadvertently standing on Steve's hand in the process. It's a good feeling knowing that the gutters are clear, and our buildings will now cope with any torrential rainstorms.

Monday, 25th June

Steve is cross today. He was out in the front and back fields, and about a hundred new molehills have appeared overnight. They're everywhere; in some patches you can't step between them and there's no grass to be seen.

Moles in moderation are good for the soil, but in this density they risk introducing harmful bacteria to the grazing, and can cause listeria in the sheep. They need controlling, but Steve and I can hammer in as many mole traps as we like and we never catch a single animal.

We used to have a local mole catcher, but he's retired, so instead we trawl the British Mole Catchers Register and find Allan, who is just up the road.

He arrives the same day and has a look around our fields.

'You've got a bad case of them,' he says, after surveying the black heaps littered all over the fields.

'I suppose you could say it's a molemaggedon,' I answer, waggling my eyebrows up and down.

He looks at me, and a long silence stretches between us.

'Right,' he says. 'I'll come this week and lay some traps, and let's see how many we catch.'

After he drives away Steve shakes his head and wanders off across the yard.

Tuesday, 26th June

Our crop specialist turned up at the farm today. Stewart (and his dogs) keeps a close eye on whatever we grow on our farm. He shows up every two weeks in his Land Rover and spaniels and walks in a zig-zag across our planted fields to look for weeds, diseases or pests.

At the first sign of meadow grass, slugs or flea beetles, he places an urgent call to Jonty, our sprayer man.

Jonty then roars down the road on his huge yellow sprayer

and green tanker and trundles around the farm applying the recommended solution. We're not an organic farm, but we're not intensive farmers either, so we don't use excessive amounts of agricultural chemicals, but we do need to protect our crops against diseases and pests.

Stewart seems well pleased with our burgeoning crop of barley, and there's no sign of any pests or black grass weeds, although he tells us he's seen a fair few crows and pigeons on the soil, scoffing the newly sown seeds.

Wednesday, 27th June

Steve is driving my little car up and down the road. There's still a strange rumbling and knocking noise in the lower gears, but he's booked it into the garage for another MOT.

After lunch we wait by the phone for the mechanic to ring, hoping that it'll squeak through the test without needing any expensive repairs. Eventually the call comes, and he explains that it's managed to pass, although he recommends that we only run it for another twelve months, in case something important drops off. I give Steve a hug. He beams from ear to ear, pleased that he's managed to fix it himself and get it back on the road. The car is held together by rust but at least it means I can putter about the countryside like before, without relying on anyone for lifts.

Thursday, 28th June

The whole family is glued to the landing window. We're watching the crows and pigeons that have landed on our barley and are stuffing themselves full of unripe grain.

Cursing, Steve rushes outside and drags the gas-powered bird scarer out from the shed, and sets it up at the side of the barley field. Our days are now punctuated with three loud bangs every thirty minutes. After a while I don't hear it any more and neither, it seems, do the birds. So we add the 'Terror Hawk' to the mix, which swings around on the end of a thirteen-foot telescopic pole. It seems to do the trick, stopping at least the pigeons from pilfering the seed.

Later that afternoon I hear the noise of galloping feet as Steve charges upstairs and reappears with his gun over one arm. He then rushes outside and shoots three times. I'm in the kitchen, and I hear him march back into the house, and go upstairs and replace the gun in the lockable cabinet.

'Crows,' he says shortly, reappearing in the kitchen. Crows, ravens and jackdaws are very clever, and it's clear that the bird scarer and pretend kite aren't dissuading them from scoffing their fill of our expensive barley seed.

The galloping feet and gunshots continue through the afternoon. By teatime Steve has shot two crows. We leave the bodies lying in the soil to warn off the other birds. Crows are clever enough to remember parts of the farm where they've been in danger, and will avoid any sign of a man with a gun.

I've read a little about the habits of crows and admire them for their intelligence. But with the exorbitant price of seed we can't afford to keep re-sowing fields just for their benefit.

Saturday, 30th June

Another wedding on the farm.

This time I'm stuck inside cooking tea, so I don't get the chance to see what the bride is wearing.

Steve has been out tinkering with his tractor and comes back for his evening meal.

'What type of dress was she wearing?' I ask.

'It was white. And sort of long,' he replies, stuffing himself with spaghetti.

'And the bridesmaids?'

There's a long silence and finally, 'I don't know ... sort of wafty, floaty kind of things,' he says while fanning his hands vaguely about in the air. He perks up a bit. 'One of the bridesmaid lasses had some amazing tattoos.' He picks up the *Hexham Courant* and takes a slurp of tea. This is about the extent of Steve's fashion knowledge.

Monday, 2nd July

I spot Scabby the ewe by the back-field gate.

She's managed to squeeze her head under the garden fence and is delicately nipping off the flowers from my border, concentrating on the clematis bush. She's grazed the lawn as far as she can reach, which is handy as she keeps the grass trimmed down around the fence posts.

Scabby is still tame, and puts in a regular appearance whenever we cut the lawn, waiting patiently for the grass clippings that are flung over the wall.

She's lost the 'toast rack' appearance from early in the year, and has filled out quite nicely. Her jaw is still undershot, so she has a constant slobber of green round her mouth, where she's been ineptly nibbling the grass and chewing her cud.

But she seems to be holding her own, and Steve reckons that we could put her to the tup in September. As long as we keep an eye on her, she should be able to raise a lamb.*

I shoo her away from the remaining flowers and sit next to her at the field gate to hand-feed her some apple slices. She loves Cox's apples, and sucks in the chunks one by one, carefully checking over my pockets for any that I might have forgotten.

When they've all been eaten she does a big gusty sigh, then ambles slowly back to the rest of the flock.

---

* The first year Scabby came to us she managed to give birth to a lovely little single lamb with no jaw deformity. This year she was far too thin and decrepit to breed from, but twelve months of good grass on our farm seems to have done the trick, and she should be fine to breed from this November.

Tuesday, 3rd July

Steve is cutting the hay field today, filling the air with the fresh smell of cut grass as the mower purrs along the meadow, leaving long rows of drying proto-hay behind the tractor.

The hay needs to dry for three to four days until we ask Jonty the contractor to charge in and bundle it up into huge round bales.

The grass is already starting to bleach in the sun and the kids run into the field to play, kicking the piles into the air and covering their knees in green stains and their clothes in tiny flecks of grass.

Wednesday, 4th July

I saw a bumblebee outside our door today. He was crawling lethargically along the porch step. I filled a spoon with a drop of water and sugar and laid it alongside him and he dipped into the sugary solution. When I checked back, he'd gone. Hopefully the extra food had given him enough energy to continue his journey to wherever he was going.

I'm less kind to wasps. They get swatted or sprayed. We do get a lot of spiders in our cottage, and I used to shriek every time I saw a big one. Now I'm less scared of them. I won't squash them, as I don't like the crispy splat noise they make, but I can manage to usher them out into the fresh air. No doubt they come straight back in again, but getting temporarily rid of them makes me feel better.

Thursday, 5th July

I've spent a happy afternoon in the Northumbrian archives at the Woodhorn Museum, looking up information about the history of our farm.

I've found a wonderfully detailed hand-drawn map dated 1825, with all the old names of our farm fields and the position of the farm house before it was modernised by Mr Clayton.

John Clayton lived at Chesters mansion in Northumberland in the nineteenth century. He was famous for saving long stretches of Hadrian's Wall as well as remodelling farms to improve their efficiency. Our farm was redesigned by Clayton in the 1840s, and it's still easy to see how the existing buildings would have been used for Victorian farming, with the brick chimney for the steam-driven threshing machine, the stables and stalls and the ranges of cattle yards with open-fronted hemmels that are so typical of Northumbrian steadings.

The map is very precise and carefully records the name of each field on the farm. They include such gems as 'Lady's Crook', 'Golden Bourie', the 'Whins', 'Lumpy South Field', 'Lumpy North Field' and 'Henry's Tack'.

Some of them are self-explanatory. The 'Lumpy' fields still look bumpy due to the rigg and furrow that curves across the surface of the pasture.

Golden Bourie refers to 'bourie' – an old word meaning a hill – and the soil in that particular field is still a bright yellow sandy colour.

A 'tack' is a field that is rented out to other farmers to over-winter their sheep. Henry must have been a long-ago farm manager who earned a little extra money from renting out the grassland. Today that field is ploughed for barley, but it would still make a well-sheltered dry pasture for the winter months.

And 'whins' is a name for tufts of long, scraggy-looking grass and gorse bushes. This is the field that we use for our horses as it's not good enough for grazing sheep.

Our wood didn't exist in the 1830s, and doesn't appear on the map, but looking at the hand-drawn tufts of grass denoting boggy ground, it's not difficult to understand why they decided to plant the wetter grazing with saplings. The trees were coppiced and provided a valuable source of firewood and income, and they still make a good wind break across the bottom of the field. The wood is called 'Angus's Covert', in memory of the farm manager who created it in the early 1900s. A 'covert' is a thicket in which game can hide. In popular memory Mr Angus was an enthusiastic member of the local hunt, so maybe he created the wood to encourage the old countryside adversary: the fox.

I pore over the map, noting the different layouts of the 'stack yard' (where they stacked up hay bales and straw), the sand pit (where they dug out the pure sandy soil for building) and the tiny circles drawn to show where the wells were dug. The streams 'Sparrow's Letch' and the 'Welton Burn' are carefully marked. The field fences in the 1830s sometimes look different to today's boundaries, and this explains why some of our field fences make a sudden jink towards the left or the right, as a leftover from this historic layout.

Friday, 6th July

Another day in the archives. I've found a photocopy of pages written in the 1890s that describe how Mr Angus, manager of High House Farm, entered and won a competition for the most efficient farm in the area. The pages describe the staff employed as cattle hands, stable lads and farmworkers, and how much they were paid (three shillings and as many potatoes as they wanted).

The farmworker cottages were 'one up and one down', and in our own house the huge fireplace where the range once stood still exists in our living room, as does the marble 'milk stone' in the larder, where they used to put milk and cream to keep it cool in the summer.

The farm buildings included 'feeding hovels', piggeries, cattle yards, stables for the many heavy horses on the farm, the dairy and the cutting edge of Victorian agricultural technology: the steam-driven threshing machine, with its tall red-brick chimney. The beautiful cobbled stall floors and the holes for the wooden partitions are still there in our machinery shed. The buildings show the low stone arches where the carts and wains lived, and you can just trace the circular shape of the gin gang behind the entrance to the brewery.

This all makes me feel much more connected to the farm-hands and shepherds who worked this land before me, and I begin to understand how a long unbroken line of men and women connect High House Farm to the past. Everything we do on the farm is done because we want to improve the land and the buildings for the next generation of farmers and farm-workers. I'm just a tiny chink in a very long chain.

Saturday, 7th July

There hasn't been any rain now for two weeks. The grass is crisping up in the heat, and some lambs in the back field squeeze through gaps in the fence to eat the longer grass at the side of the barley fields.

Steve is worried.

'Four months ago we were knee-deep in snow with no bloody grass, and now everything is burning up and there's still no bloody grass! What's next? Rivers of blood? Plagues of locusts?'

It's been the most extraordinary year weather-wise. It makes me long for the usual intermittent sunshine and regular grey drizzle that normally makes up a Northumbrian summer.

Wednesday, 11th July

Early this morning I spot a dead lamb lying in the entrance of the front-field gate, and for the life of me I can't work out what killed it.* It doesn't seem to have been attacked by a fox,

---

* Every year we lose one or two lambs to disease or predators – all farmers are the same, and it's recommended to budget for losing 1–2 per cent of our lamb flock.

and anyway the lambs are too big now to be taken by a hungry animal. There's no obvious blood on the carcass. We pull it onto the bike and I see that it has mucus running from its mouth. It must be pneumonia or 'watery mouth'.

We bag up the dead lamb in a sheep feed bag and stow it in the corner of the old cattle pens. I walk into the field to find the mother and see if her other lamb looks poorly. She's marked as number '25' and she's lying in a dip in the ground, her lamb tucked up next to her fleece. Catching him is another matter, but I manage to hook the curved end of my crook round his neck and pull him in for a check. He doesn't have any snot running from his nose, but we can hear the slightly laboured breathing that might be the beginnings of pneumonia. Steve injects him with an antibiotic and we shoo both lamb and ewe out of the field and into the warmth of a shed.

Pneumonia isn't always a disease of cold and driving rain. Often, damp and warm weather for a couple of days can bring on the infectious disease. Hopefully by removing the pair from the field we'll stop the spread of the illness and this lamb will survive.

Thursday, 12th July

The next morning the poorly lamb looks better, but there's still a faint 'ruttle' to his breathing. His mother is having the time of her life, having access to ad-lib silage and a regular delivery of sheep feed. We'll keep an eye on them both, but they're better under shelter.

There's a wedding at the brewery today, and all morning there's been a steady flow of cake decorators, flower arrangers and musicians turning up to decorate the site and set up their instruments.

The wedding ceremony is taking place in the hay barn, which has been decorated with a pink and cream flower arch and swags of beautiful roses from corner to corner. As always,

Heather presides over the arrangements with an aura of tranquillity and calm. She's been here since the early hours, and will stay right until the party finishes. Her staff are dressed in crisp black shirts and dark green aprons as they lay tables, set up the hog roast for the evening and make final adjustments to the decorations.

We lean over our garden wall to watch the arrival of the guests. They come in a big red London bus. We've seen plenty of different wedding transport over the years, from a vintage combine harvester that struggled to make the turn into the farmyard, up to sleek Bentleys and cute antique VW camper vans.

Suddenly, a big grey lorry appears around the corner, with vents cut into the top.

'Oh hell,' says Steve, 'it's the knacker van.'

He's forgotten that he's organised for our local Fallen Stock collection company to pick up the dead lamb that had pneumonia. By law, we need to ensure the collection is done within twenty-four hours.

The knacker company are incredibly efficient, but this spring has been so dire that they must have been rushed off their feet. The van screeches into the farmyard, scattering the guests, before reversing in a hurry up to the sheds.

Unfortunately, it's quite a warm day, and the smell issuing from the vents at the top of the lorry isn't very wholesome. The bride has already arrived on the arm of her father, and as she steps out the car her nose wrinkles at the sudden whiff of decomposing sheep.

'Sorry! Sorry!' shouts Steve as he charges past the marquee. Heather watches him with narrowed eyes as he instructs the driver to reverse into the sheds at the top of the yard. He waves at her apologetically. Shaking her head, she retreats into the brewery, ready to welcome the bride and her father.

Ben, who loves the knacker van visits, rushes to pull on his wellies so he can see what dead animals are inside. The driver unlocks the back door and pulls down the ramp to reveal a sad heap of dead ewes, a limp Simmental heifer and a veritable mountain of lambs that the van has collected on its headlong trip around Northumberland.

The driver pulls a sheaf of papers out of his pocket, peels off the top sheet and thrusts it at Steve. The poor dead lamb is added to the top of the heap inside, and once the doors are closed the driver hops back in and hurtles back out the yard.

'Don't go through the ...' starts Steve. 'Oh. He has.'

The smell of dead animal follows the van as it races through the yard full of guests, scattering them to both sides. We see smartly dressed visitors wafting wedding menus to try and dissipate the smell.

We'll have to apologise properly to Heather about the interruption to her day. Normally we keep as far away as possible from guests, as a couple of scruffy farmers isn't really a great addition to any wedding.

You can't blame the knacker man either. He's under pressure to pick up fallen stock within a certain time frame, and normal farms don't tend to hold big society weddings in their front yards.

Ben is bubbling over about all the dead animals he's seen.

'There was a big red cow. Did you see it Mum?' he says. 'And did you see all those lambs? He didn't use the winch this time though,' he says, looking disappointed. He loves it when the knacker driver uses the long chain to haul a big beast into the back of his lorry.

Saturday, 14th July

Candy is in disgrace. Again. She spent yesterday rolling in the run-off from the septic tank, and is covered in black, stinking mud from her ears to her hooves. She was very pleased with herself, and happily marched into her stable, while everyone recoiled from the stench.

Monday, 16th July

Steve is filling the fertiliser spreader with a mix of white phosphorus and potassium grains. It's a dry day and there's no wind, which is ideal weather to spread the little fertiliser granules onto our grass fields. We use a mix of the two chemicals instead of straight ammonia, as ammonia is too strong for our land and tends to 'sicken' the ground.

I have a less exciting job. I'm mucking out a month's worth of manure from the pet lamb shed. It's too small and awkward a space to use the loader and scoop, so I have no choice but to do it by hand. After half an hour I'm wheezing like a pair of bellows and absolutely dripping in sweat.

Dad turns up for a chat with a thermos full of tea.

'You're holding the fork wrong,' he says, looking at me over the rim of his cup.

'Bugger off,' I say quietly.

'Hold the fork handle with your left hand on the top, not underneath. Otherwise you'll give yourself a rupture.'

'Oh, for God's sake, Dad! I'm knackered. You have a go.'

I thrust the rusted fork at him, and he sets to, heaving up great forkfuls of smelly straw and sheep poo.

Between us, we manage to clear the shed of wet straw and pile it all into the bucket of the scoop, ready to tip on the manure pile down the road.

I sit on an old tyre and share another cup of thermos tea with him, while watching a starling flit in and out of the barn, catching flies to stuff into her nest of babies.

I could sit like this for hours. Dad and I talk about everything, from the Battle of Britain to the walks we used to take around the hills of Cheviot. Days like this are absolutely golden.

Tuesday, 17th July

The pneumonia lamb and his mother look much better, and we turf them out in the sunshine of the paddock. His mother takes some persuading to leave the lovely warm shed, but once she tastes the grass she doesn't lift her head for the next hour, while her baby skips around her and eventually lies down to sleep on a warm molehill.

Allan's mole traps seem to have done the job. He's caught an incredible total of thirty-eight little beasties. As tradition dictates, he carefully shows us the pile of bodies, so that we know he's telling the truth and not trying to declare more moles than he actually trapped.

To people unused to country life, it looks like a massacre, but culling of any out-of-control species is essential to ensure the health of the carefully balanced ecosystem of our fields, wildlife and animals. In the old days mole men used to hang up each tiny animal on the fence as an advertisement, and to show off their skill. They don't do this any more, and I'm glad, as I don't like seeing the desiccated corpses swaying in the wind.

Like all well-run farms, we have a company that comes out to deal with any other pests, such as mice and rats. They place little trays of poisoned bait out to stop infestations, which could cause disease.

Having said that, yesterday, when forking through that pile of manure, a rat shot vertically out of the straw and scuttered off up the wall into the rafters, almost giving me a heart attack. They move so fast they make me shudder. I don't mind tame rats; in fact after I've watched a few videos online they seem to me to be clever little animals. Wild rats are something different though. I make a mental note for Steve to contact our pest control guy again. Maybe we need some more traps put down.

Wednesday, 18th July

This week it's been warm and bright, and today I take a minute to sit on our bedroom windowsill and watch the birds outside.

I hear the falling liquid notes of a blackbird, the metallic 'chink, chink' of chaffinches, and the good-natured wheezy squawks of a starling that has made its nest under the eaves of our roof. Our farm hedges are carefully trimmed so they grow thick and sturdy, and the blue tits love to nest in them. They fight and chatter and fly down to dust-bathe in the cracked mud at the base of the branches.

I hear the velvety 'coo-coo' of a wood pigeon and the raspy 'chuck, chuck' from jackdaws gathering in the old ash tree at the bottom of the front field. There's a flash of yellow as a yellowhammer flits into the pine trees lining the road, and everywhere you look there's a babble of sparrows, gossiping and arguing and flitting from branch to branch.

We used to have red squirrels that lived in the pine trees, but over the past five years they've died out. We see a lot of grey squirrels scampering up and down the bark, and they grow bold enough to visit the bird feeders we put out over the winter.

All the birds are busy and preoccupied, either building nests or feeding young. From the window I see a solitary heron standing, one leg tucked, next to the Sparrow's Letch at the bottom of the front field. He lives at the nearby reservoir but likes to visit our farm to feed on the minnows in the stream. Suddenly he takes off, legs trailing behind him as he lumbers into the sky.

Thursday, 19th July

It's shearing day. It's hot and humid when John the shearer turns up, driving his truck and trailer into the yard.

John sets up a large folding wooden platform that looks a bit like a stage. The sheep are gathered into the shed, then shooed, one at a time, up a sheep race to the top of the platform. One sheep is penned in at the top to tempt the others up – sheepy psychology!

Mrs Snuff the Suffolk is the sheep at the top, and she seems quite comfy – she lies down and baas continually at everyone throughout the whole of the shearing. We feed her sheep nuts to keep her happy.

The rest of the sheep run up towards Mrs Snuff and John grabs them one at a time through a trapdoor set into the side of the platform, and sets them on the wooden boards ready to be sheared.

He sticks some loud dance music on and starts. It's amazing watching him – he wears moccasin slippers so he can grip on the wooden floor, and balances the sheep on their left hip on top of his shoes so they can't get traction on the ground to twist away. The first cut of the wool is down the belly, then you 'open out' the fleece from the back leg, right up to the neck so that it falls cleanly away, and then up and over the back and finish at the rump. It's so smooth and fast. When John's finished with a ewe it looks so clean, and its cropped fleece is almost blindingly white.

He drinks a four-litre tub of water every hour to keep hydrated, as it's such hard work. I want to have a go, but can't get the hang of positioning them on one hip bone.

Fatty the sheep gets stuck in the crush. She's so fat she's shaped exactly like a rugby ball, but we discovered a lot of the roundness comes from her thick and springy fleece. Once sheared she looks like all the rest – a slightly startled but very clean goat sheep shape.

The lambs aren't sure which sheep is their mum without their coats on – so there is a lot of shouting and baa-ing and general mayhem until they are sorted out. You can see the lambs sniffing each ewe that comes off the platform until they find 'their' mum, and then they trot off quite happily.

Dad and I wrap the fleeces, and this year Lucy helps to catch the fleeces and pack them tightly to send to the Wool Marketing

Board. The Wool Board 'sheet' is like a massive plastic pillow-case that is open along one side. When each sheet is jammed full of wrapped fleeces I take a huge curved needle threaded with baler twine and sew the sheets tightly closed. We then mark on each sheet how many fleeces it contains and whether they come from hoggs or ewes (hogg fleeces command higher prices as they're from an animal that has never been sheared before, and the wool is softer).

We get around £1.60 a fleece from the Board. But of course we need to pay John to shear each sheep, so we hardly break even. In the 'old days' it was said you could pay your yearly rent with your wool payments. Not any more. Some farmers are selling direct to the customer – to weavers and spinners who work with the raw wool. But there just isn't the demand for our rougher Texel and Mule fleeces. It's a pity, as it's a gorgeous product. When the fleece has just been sheared it's thick and heavy with lanolin (a natural type of waterproof oil, often found in soap), but once washed, carded, brushed and spun the wool is soft as silk.

It would make more financial sense to shear the sheep our-selves, but Steve has knackered his back, and I just don't have the strength. I want to learn though. I wonder if you can shear a sheep that is standing up and tied to a fence?

Friday, 20th July

I'm watching our flock out in the back field, and the ewes look so much happier without their heavy fleece. There's now no risk of fly strike or maggots, and they're much cooler and less itchy. Mabel still has her taupe-coloured legs and face, and stands out against the startling white of her sheared wool. Spotty Nose is saggy all over and has lopsided udders. She's the oldest in the flock though, so you can excuse a bit of slippage in her old age.

Saturday, 21st July

The school holidays are looming. When the kids were younger I found the holidays hard work, as they needed to be watched like a hawk, and trying to do that *plus* look after the farm and do office work was exhausting.

Now they're older it's much easier, as they wander off into the farm and entertain themselves. They're sensible enough not to fall off hay bales or fall in the river or talk to any strange walkers on the public footpath. I kit Lucy out with a backpack containing her phone, a toilet roll (through bitter experience I've learnt that children always need a poo when miles away from the toilet), snacks, a drink and some sun cream. She dances off with Ben trailing behind her to do the things they love doing: building camps, climbing trees, fishing in the stream with a net, making secret treasure hunts and sitting with the pet lambs in the sun.

Two hours later they're back. Ben has fallen in the stream and is squelching with every step. Lucy has stepped in some nettles and has stings up the back of her legs. But they're very happy. I stick them in the bath and scrub off the mud, and then it's pizza and chips for tea.

It sounds idyllic, and of course I hope the kids will look back when they're older and see how lucky they were. At least I can give them the alternative of being outside and the freedom of having their own time and space in the fields and wood at High House.

Sunday, 22nd July

The lambs are having races in the back field. I can see them while I'm sitting at my desk. They gather in a group under the beech trees, and then suddenly one dashes off to the top of the hill and the others stream behind him, bucking and jumping into the air. On cue they turn and happily race down the side of the hill right to the bottom, past their grazing mothers, and then curve around to gallop back up to the top again. They like to play in the evening or first thing in the morning. Even Titchy the tiny lamb is joining in. He's half the size of his bigger flockmates, but he still races along as fast as he can, his tiny woolly legs a blur.

The lambs are also 'stotting', or jumping with all four feet, off the ground along the hard soil at the edge of the field. It's a move peculiar to sheep and goats, and we have a theory that they do it on hard ground as they like to hear the 'tock, tock' sound their hooves make against the packed earth.

After they grow tired of racing, the little ones wander off to play 'King of the Castle' on the big stones in one corner. One lamb jumps up onto a rock, and the others try to push him off, lowering their heads and charging until he's forced to jump to one side while another takes his place. The game goes on until one of the ewes raises her head to call for her lamb, which races across to duck its head under her belly, thumping upwards into her udders so that she releases her milk.

Monday, 23rd July

This morning the first caravan of the year pulls up in our farmyard. I wander over to say hello. They're a nice elderly couple from Yorkshire. The wife fries bacon on a camping stove while her husband asks me about the trees, birds and animals, and we have a good chat over the field gate. Mid-conversation, he suddenly pulls out a banjo from the back of his van and starts serenading a bemused fat pony.

We're a member of a clever scheme called Brit Stops, where motorhomes or caravans can stop for free in our brewery car park so long as they buy breakfast or lunch from the tearoom. We get a lot of German, Austrian and Dutch people, who almost always speak perfect English.

I'm not sure I can cope with banjos this early in the morning, so I make my excuses to Yorkshire man and leave him singing away to Candy.

Wednesday, 25th July

Summer is in full force.

It's so hot I'm lying on the lounge carpet with a wet tea towel on my head. Steve is repairing the back-field fence in the blazing sunshine – I can hear him thwacking in posts and grumbling about being dehydrated through lack of tea.

The pet lambs have overcome their agoraphobia and are now definitely enjoying their new freedom in the paddock. If you sit down on the grass, they like to come and chew their cud and have a neck scratch.

Fuzzy the lamb stretches out his chin so you can stroke behind his ears, and burps happily while staring lovingly into your eyes. The lambs fall asleep on you if you're prepared to sit there for a while. It's cute when they're tiny but not so lovely when they grow into huge woolly ewes that still demand to sit on your lap.

I often have lunch outside. The avian flu restriction has lifted, and Marjorie and Ethel are now mooching happily round the farm. Marjorie the chicken can spot someone eating a sandwich from miles away. She extends her neck like a periscope, sees my sarnie and belts over, clucking frantically. She stomps around in a circle and pecks at my feet until I give in and feed her a crust or two.

Marjorie's favourite thing to do is join in on weddings, either by strutting down the aisle in front of the bride and groom or by annexing the play area and terrorising small bridesmaids into feeding her crisps all evening. If you go to pick her up she crouches down shivering, pretending to be all frightened, or slips under the decking where no one can reach. She's a law unto herself. No one has complained yet, but I reckon there must be many wedding photos out there with photobombing Marjorie posing for the camera.

Thursday, 26th July

Still no rain. The streams are still running, but only just, and the ewes look thirsty. We open the fence between our front field and the horse field so that the sheep can drink from the mains-plumbed trough.

Candy watches grumpily as a long line of ewes and lambs march through her paddock to drink from her water trough. The flies are legion, and she's started scrubbing her bum and neck against the wall, rubbing her skin raw in patches. I shove her inside her cool stable and she almost immediately falls asleep, grateful to be out of the hot sun and incessant biting flies.

Friday, 27th July

I walk over the paddock and the grass feels crispy beneath my feet. It's so blasted by the sun and lack of rain that already big sections have gone brown and died back. We only have one bale of hay left until the new crop is cut, but Steve decides that we'd better feed it to the ewes and lambs as the grass is so poor.

He digs out a huge circular ring feeder, sets it down in the field and fills it with our last precious bale of hay.

The ewes don't take a lot of interest, but at least they have extra feed if they need it. We can't afford to buy in extra forage, as the poor spring means there's a UK-wide shortage of hay and straw, and what little available costs a fortune.

The troughs are still running, but we can hear birds scratching in the gutters of the farm buildings, a thing they only do when it's very dry. They must be looking for leftover pockets of moisture in the dead leaves.

Steve uses the pressure washer to make a big puddle of water in the yard. This will give the birds a drink, and they'll be able to have a bath. Swallows immediately fly down and start digging in the muddy pool, and when we leave a female blackbird is having a good old splash and wallow in the water.

Saturday, 28th July

This Saturday the brewery is hosting a pagan wedding.

I sit in the front garden to watch the bridal party drive past.

They've already asked our permission to go into our wood after their handfasting ceremony for a 'personal ritual' with an ash and oak tree. Which is fine by me. Communing with trees is high up on my list of 'nice things to do', so this rather lovely young couple is free to go and hug a tree, or whatever it is they're planning.

The bride wears a circlet of fresh flowers and a long, flowing dress, and the groom wears a baggy white shirt and has a red handkerchief tied round his neck, making him look rather like he's about to take part in morris dancing. After the civil ceremony the whole wedding party, including a priestess-type person in a long red cloak, marches down to the wood to carry on the celebrations.

The after-wedding party seems to go with a swing, and there's a hog roast and a ceilidh in a marquee afterwards. We creep around the tents, checking on sheep and lambs while the drums and fiddle play on into the night.

Sunday, 29th July

I'm out at the sheep pens the next day. We've kept a handful of ewes and lambs that need a little extra care and attention in one of our smaller sheds. When I go and do my morning check I find them in the farmyard, munching happily on the grass that pokes up through the cobbles. The door to the shed is standing wide open.

One of the wedding party must have ignored the clear 'NO ENTRY' signs and gone into the shed, then forgotten or been unable to close the gates behind them.

Was it deliberate? We decide that it was probably a couple trying to find a dark place for some after-hours shenanigans. I don't think the sheep shed would be a very romantic place for any illicit goings-on, as our sheep would probably stand round in a tight circle, breathing heavily and watching you extra carefully.

I shoo the ewes and lambs back into the shed and push the gate bolt back into place.

Tuesday, 31st July

A couple of months ago we agreed to a Caravan and Camping Club from down south who wanted to spend a weekend at our farm.

It's been organised by a couple of ladies, and today they've turned up with lots of friends and family. As we have no facilities like showers or toilets or even electricity they have hired a porta-shower and three porta-toilets.

They'd expected 140 people to camp, but we counted at least 250 that drove in on the day. They've camped in the paddock next to the brewery and when the kids start running around and the music cranks up it looks like a small, happy festival.

They seem organised, and one of the ladies has a big list of camping rules on the campsite (no fires, no loose dogs). And there are proper recycling bins, dog-poo toilets, pretty tents, fluttering flags and whatnot.

I watch a parade stomp around the field, with people banging drums and singing and everyone waving rainbow banners. It all looks good fun.

Eventually, I catch a reluctant Candy and haul her into the camping field so that the kids can pat her and ask questions. Many haven't been on a farm before and are really interested in the sheep and chickens and crops and want to know why the chickens are peering into their tents, and if they can take a pet lamb back home.

But there are lots and lots of kids, and without being all judgemental, I don't think the campers realise that a farm isn't somewhere you can just let your kids run free like the wind, while you drink two litres of cider and get hammered in your tent.

I find a group of enormous teenagers playing football and bouncing the ball off the ancient drystone walls, which knocks out huge great chunks of stone.

Then I see a little kid perched on top of one of the hemmel roofs – and I go completely nuclear. Steve said I'm like a very small, very angry tornado. The roofs are *not* safe – and if he had fallen through and hurt himself can you imagine the health and safety nightmare? We'd be tarred and feathered and paraded through Hexham.

I scream at the kids, 'If I catch you little sods on top of the roof again you will be going home tonight!' and stomp off to tell the parents.

'Oh dear,' they say, while drinking cans of lager and burping happily.

I've gone to bed with a headache, accompanied by sounds of banging drums and screaming kids.

Wednesday, 1st August

Today Steve has discovered that someone has jumped up and down on one of the metal gates and snapped a huge stone gate-post right the way through. The post was a foot wide, dates from 1840 and costs at least a thousand pounds to replace.

The kids have also been sneaking into the horse field, chasing round an alarmed Candy and daring each other to grab the electric fence, thereby shorting out the whole system and draining the battery.

I went, po-faced and furious, to tell the organisers who were terribly apologetic – but of course don't have the money to pay for a beautiful 170-year-old gate post.

So that's the end of the camping, and after a short break-fast, everyone starts wending their way home, cars loaded up to the gunnels with hastily pulled-up tents and sleeping bags wrapped haphazardly around wailing children.

One amusing postscript comes from the poor porta-potty man who had to come twice to empty the loos during the weekend. He said they should have had at least six toilets to cope with the number of people using them, and he'd 'never been twice over one weekend to empty them before'. He was kitted out in every hygenic protection device known to mankind, but even he reeled with the smell when he opened the toilet doors.

Thursday, 2nd August

I have the fattest horse in Northumberland. The farrier came to trim her feet, and Candy is so embarrassingly rotund that she can't lift her front foot past knee height. (I have the same problem.) Joe my lovely (and rather handsome) blacksmith managed to winch her knee high enough to do her feet, but it was a bit mortifying. However, Candy could not give a tiny stuff.

She's been belly-rolling under the electric fence to get at the good grass in the next field and has been discovered ears-deep in a sheep lick in the adjoining paddock. I'm paranoid about laminitis (a condition that small, fat ponies are prone to in the summer) so she's now on a strict diet and has hardly any grass at all – just barley straw to eat.

What she needs is a couple hours of work every day, but down to kids, work and everything else, Candy hasn't been ridden since May. She hangs sadly over her stable door, a strand of straw drooping from her lips, and whinnies pathetically at anyone who walks past.

Friday, 3rd August

I take a walk down into the wheat and barley fields. The plant stalks are swaying in the wind and the fields look like rippling oceans of green under the hot sunshine. It won't be long before they start to ripen in this weather.

I've learnt to tell the difference between the crops: the barley stems are a slightly lighter green, and the plants have long feathery spikes (awns) that stick up from the hard grain. The wheat plants are shorter, darker green and don't have any bristles.

The oilseed flowers have disappeared to leave a thick carpet of greyish plants, and the entire field now smells strongly of cabbage, as the seeds ripen ready for harvest. Visitors are already delicately mentioning the 'bad smell' when they come to the house, and I make a mental note to ask Steve if he can plant next year's oilseed crop in a different field, away from our kitchen window.

Saturday, 4th August

Another party of campers have come to set up tents at the farm. They're staying over for a wedding on the Saturday night.

They seem to be a nice group of lads; I wander over to tell them about shutting the gates and apologise for the amount of sheep poo that's in the field.

Everyone is very pleasant, and no one seems to mind Marjorie the chicken investigating each of their tents in turn. She carefully pecks through their things, searching under sleeping bags and poking her head into each tent. The pet lambs in the paddock are just as irritating, marching up to snuffle through the campers' belongings. I remind everyone to keep their tents closed to prevent the animals moving in, climbing into sleeping bags and eating their food.

The next day I zoom past the paddock on the quad bike

and find the pet lambs sunning themselves at the bottom of the paddock. Fuzzy seems to have a frill round his middle, and getting closer I realise that he's managed to put his head into a plastic bag and is wearing it like a plastic tutu. The lamb looks quite proud of his stylish accessory, and while I disentangle him I search around the grass for any more rubbish. Apart from a few empty beer cans piled neatly in a bucket, there's nothing at all. I like these campers.

They emerge, all bleary-eyed, about 10 a.m. I give them some fresh eggs from Marjorie and Ethel's nesting box to apologise for the fact that the chickens have spent the hours since daylight poking their head under the tents and clucking loudly at the inhabitants.

I must remember next time to try and keep the chickens out of the camping paddock. I can just see them climbing into someone's car and being bundled off by accident.

Monday, 6th August

I was walking through the fat pony's field when I stubbed my foot against something in the grass. Pulling it out I saw that it was a shard of grey pottery with the words 'Numol – Tonic and Nervine Food for Children and Adults'. There were a few more pottery flakes lying on the surface of the ground.

Maybe the past inhabitants of the farm threw their rubbish over the wall into the field? The land certainly rose up sharply at the point where I found the pottery, and the dry weather has made the soil so crumbly that it's easy to spot anything poking out of the ground.

Calling the kids and dragging some gloves and spades we start excavating the area. Almost immediately my spade clinks against something hard, and I pull out a cream and brown jar. This is exciting; we start digging like demons. After an hour we've amassed a small pile of china, glass and pottery fragments. There are old green and blue bottles, a brown earthenware teapot, a tiny glass perfume bottle marked 'Lavender', some pretty blue-and-white china pieces and, best of all, a couple of stoneware jam pots stamped 'Hartley's Finest Marmalade'.

There's also a huge collection of old batteries, bits of miscellaneous wire and metal, and huge pieces of pottery animal troughs.

This is brilliant. The kids and I feel like proper archaeologists. We must be digging through the old farm midden, where all the old broken plates and jars and bottles were thrown. We wash all the fragments in the horse trough and lay them out on the kitchen table. I spend a happy evening researching all the bits and find out that the jam jars are late 1880s and the Numol pot is from the 1920s.

'Be careful you don't stab yourself with something,' says Steve when I show him my haul. 'There's some wicked pieces of old glass and iron in there.'

I make a mental note to buy the kids reinforced gloves and check that their tetanus injections are up to date ...

Tuesday, 7th August

I'm just not used to the heat. Steve and I crouch in our house every morning, slapping sun cream on our wizened, sun-baked bodies.

As Northumbrians, summer usually means a low-grade mix of rain, the odd spot of sunshine and howling winds. Constant sunshine is unnerving.

Steve has broken out his shorts. I check the sheep in a terrible floaty kaftan I've found in the back of the wardrobe. Only the kids are unaffected, and splash happily in the bug-infested paddling pool I've set up in the garden. They're already brown as berries, and their hair has turned into spun gold.

Wednesday, 8th August

Worming day. The entire flock needs to be brought in so that the ewes can be given a wormer plus a booster of minerals and vitamins, ready for the tupping in November. We start early to beat the heat, but it's still stifling – very humid and muggy.

Poor Mavis is flagging in the sun, so we give her a drink and put her into Candy's stable to cool down. We've seen collie dogs go into convulsions with heatstroke when the weather is hot.

I take her place and charge about the field, flapping my arms and uttering squeaky barks to try and move the flock up through the gate. The ewes are slow and sluggish, not wanting to move in the scorching sunshine. Eventually, Steve and I steer them up and through the gate and into the cool of the shed. I'm absolutely lathered with sweat and covered with tiny black harvest flies that look like flecks of dust and make me itch all over.

Each sheep is given a dose of wormer and a second measure of vitamin and minerals. Mabel is always first in the queue, checking my pockets for sheep nuts and demanding a neck rub. The older sheep are easier to handle, as they know what to expect and stand fairly patiently as we squirt wormer into their mouths. However, some of the younger ewes are fairly bonkers, throwing themselves against the metal bars of the pens while we struggle to hold them down to push the wormer gun in-between their jaws.

Finally we're finished, and we lead the flock back to their field: Mabel and Spotty Nose in front, and Fatty bringing up the rear. Mavis emerges from her stable, yawns and trots along behind, stopping every now and then to investigate something interesting in the hedgerows before bounding up to Steve and me, tail wagging.

I go home, peel off my sheep-shit-encrusted leggings and stand under a cold shower until my body temperature reaches normal levels.

Thursday, 9th August

My lovely mum is 72 years old today. We've organised a family barbecue in our shearing shed to celebrate. It's still full of sheep dung and clipped wool, but we pull out buckets to sit on, stick some sausages on a cheap disposable barbecue and sit back in the shade.

Lucy and Ben ride their bikes up and down the empty shed while Mum sits on a sheep lick bucket and carefully disentangles a thread of wool from her hot dog. Steve has made a sponge cake with strawberries and cream, and after belting out an enthusiastic Happy Birthday, we sit with disposable plates and spoons, scraping up the cream and swigging cups of tepid tea.

We've decided to camp out in the front paddock as an extra birthday celebration. We've got two tents – a one-man pop-up shelter and a much bigger round bell tent that I've borrowed from a friend.

As dark falls the bats flit in and out of the shadows, and we sit round a camp fire, all talking at once, laughing and joking and telling stories.

Dad has had a bottle of beer and is telling Steve the story (now in family folklore) of when he went to Iran in '76 and had such bad diarrhoea he had to do a poo in the desert.

'The trick is,' he says, waving his bottle around, 'to kick it about in the sand a bit until it all disintegrates.' Ben's eyes are like saucers.

The children are terribly excited at the fact that we're all going to camp, and they run in and out of the tents, squeaking with excitement.

Steve commandeers the one-man shelter and happily settles down in his sleeping bag. The rest of us, Mum, Dad, Lucy, Ben and I, are on sleeping mats arranged star fashion around the central pole in the big, echoey space of the bell tent.

The kids quickly settle down, scooting into their thick sleeping bags. I've layered an extra double duvet on top.

Mum's sleeping bag seems very tight.

'I can't seem to turn around,' she says faintly, with the draw-string hood pulled firmly around her head and down to her nose. All I can see is a tuft of grey hair protruding from the top of the bag.

Dad is grumbling that he's lying on the thinnest ground mat in the world, but before long he falls fast asleep, his snores echoing inside the dusty interior of the tent.

Night draws on and I am absolutely frozen. I've never been so cold. My ground mat is bumpy and my sleeping bag is too thin. Everyone else is now peacefully asleep but I can't get to sleep as my feet are like blocks of ice.

I try wrapping my feet in a black bin liner to keep warm but that doesn't seem to help, and I just rustle every time I turn over. I now need the toilet, so I tiptoe out of the tent in the pitch black, fall over a chair that someone has left in the path, and accidentally wee on a guy rope.

Back inside I fall into a fitful sleep, huddling down next to a peacefully breathing Lucy, while trying to tuck my legs under her layers of duvet.

Friday, 10th August

I woke up stiff and cold this morning. Dad is already awake and begins to disentangle himself from the duvet, slowly haul-ing himself from his prone position, complaining about a bad back and a sore neck.

Ben and Lucy are as fresh as daisies. Mum's sleeping bag had been so tight that she hadn't been able to turn over all night, and her hair has swirled itself into a point on top of her head, like an ice cream cone.

Steve bounces out of his tent, rubbing his hands together, 'I slept really well last night!' he says, unpacking the breakfast things and fiddling with the paraffin stove to get it to light.

I ignore him. I feel damp and grubby and have a bright red

nose. Mum and I sit on either side of the stove, her surreptitiously rubbing the marks made by an overtight sleeping bag and me rubbing my feet to get the circulation going.

The next time I camp I want it to be on some luxurious tented safari, with butler service and hot and cold running water.

Sunday, 12th August

We're in the front field counting our sheep. Mavis doesn't like sitting out in the hot sunshine, so she carefully makes her way over to the water trough, jumps over the edge and lowers herself into it with a sigh. All you can see is a pair of brown eyes and a black nose and her curly tail poking out of the water like a flag. Steve whistles and she splashes out, shaking her whole body, throwing a shower of water droplets over us and any nearby sheep.

Monday, 13th August

I'm mooching through our wood, absent-mindedly swinging my thumb stick at the brambles on the side of the path. I'm here to see if the stream is still running.

Welton Burn has slowed to a tiny trickle, and the left-hand stream branch where it splits around the base of an oak tree is completely dry. I walk along the dry riverbed, slipping a little over the damp rocks and stepping over all the dead branches and twigs littering the bottom. A dead ash tree has fallen across this section of the stream, and I sit down to watch a teeny bank vole scurry along the wet soil at the base. He has a very white tummy, a brown furry back and long luxuriant whiskers. He stops occasionally and sits on his hindquarters to clean his face with his two front paws. He's on the hunt for worms or beetles.

The brambles and nettles are too high to walk back on top

of the bank so I return via the oak tree along the damp pebbles and rocks. Underneath the tree there's a well-trodden path over the riverbank where animals have been drinking. I spot a badger print (five distinct toes and claws) in the mud. Their sett is close by, and I wonder how they're managing to dig for food in the dry ground.

As I walk back to the farm the sky miraculously clouds over, and I can feel a few spots of rain on my face. I rush into the house to tell Steve, and we watch the rain spots on the window, hoping and praying that we get a good soaking.

Tuesday, 14th August

We've finally had some rain. Last night the skies opened with an almighty deluge of thunder, lightning and water. It rained solidly for most of the night. This morning the ground is steaming, and it's feeling cooler.

Already the paddocks are beginning to look less dusty and brown, and our back lawn has started to green up.

Wednesday, 15th August

Lucy and Ben are busy setting up a roadside stall to sell some of our surplus eggs and vegetables from the allotment. It's a good way for them to rustle up extra pocket money, which no doubt they will splurge on the tombola at the next village fete.

They haul out a badly spelt blackboard ('Eggs and Veggitabels') and pile a wonky table with big leeks, huge potatoes and boxes of brown eggs, freshly laid by Marjorie and Ethel. They then spend a good ten minutes arguing about the prices, until I intervene to say that £10 for six eggs might be pushing the market a bit.

I keep a careful eye on them out of the kitchen window. I can see passers-by stopping in front of their stall, and after an hour of frenzied selling, a breathless Lucy rushes in to show me their profit. A whole six pounds! They were even given an extra pound each by a member of the brewery staff who felt sorry for them. A great result. Ben is already calculating how many Pokemon cards he can buy, while the sensible Lucy carefully adds the coins to her piggy bank. It's already too heavy to pick up, and she tells me she's saving for new drawing pencils and paints.

Thursday, 16th August

Today is a big sheep day, as we're weaning the lambs. We call it 'spayning' on our farm, and it usually heralds a day and night of constant bleating as we separate the ewes and their offspring.

The lambs are now fully grown and big enough to manage on their own without their mother's milk. Our ewes need a few months off, without their grown lambs, to build condition until they meet the tup again in November.

We separate the lambs from their mothers by running the whole flock through a sheep race with two gates at one end. The ewes go into one pen and the lambs go into another. Then we herd them into two fields, one pasture at the east end of the farm, and one pasture at the west.

The ewes occasionally lift their heads to give a half-hearted baa, but seem grateful to be separated from their enormous children. Over the past few weeks I've noticed that the lambs lift their mothers off their back feet when they try to tuck underneath her to reach her milk. They kneel into the grass and thump upwards with their hard heads, butting into her udder. It must be painful, and the ewe usually marches across the pasture, walking over the lambs and shaking herself irritably.

Our lambs are getting nice and fat, almost ready to go to the Mart. Steve walks round the pen, checking how ready they are by feeling their plump little tails – and some are almost round, especially Fatty the ewe's two offspring, whose little woolly tails are like fluffy tennis balls.

'Nooooooooooo,' says Lucy, when I mention the Mart, 'can't we keep them all?' But we really can't, as we'd have elderly sheep staggering around the farm like the living dead, and our lambs give us a good chance to make a decent profit this year.

In an ideal world, once we've separated the ewes and lambs, they blare and baa on for a bit, and then they get used to it, and the mothers think 'Thank god for that', eat lots of grass and start to put some condition on, ready to be tupped again in November by Randy Jackhammer and Thrusty Clappernuts.

However, once all the flock is separated into their two fields, and we've gone back to the house for a cup of tea, one lamb manages to wriggle under the gate of its field, belt up the farm drive and press itself against the field gate right next to the ewes. This gives the rest of the lambs an idea, and by this afternoon we have twenty-five lambs out of their field, ricocheting around the farmyard and trying to squeeze themselves through a two-inch gap in the fence to find their mothers. Who are all completely unconcerned and are grazing determinedly with their backsides towards their bawling children.

We try to gather the lambs and herd them down the drive back into their field. It's impossible. As soon as we shoo a few away from the gate, they double back and go to mad lengths trying to break through the gate, even throwing themselves at the fence at head height.

Lucy goes off in a huff after Steve shouts, 'For god's sake stand still!' at her too many times, and Ben sensibly refuses to get off the quad bike and watches us while munching on a packet of salt-and-vinegar crisps. Mavis is hiding under the trailer, beadily staring at the lambs, but refusing to join in after being shouted at for nipping a back leg.

Eventually in frustration we manage to push the lambs back into the sheep race, shove them into the sheep trailer and then drive the few hundred yards to their field, heave them out of the trailer and into their field and stick a big log in front of their gate so they can't wriggle out any more.

Friday, 17th August

The lambs were shouting right through the night but finally tailed off early this morning. We blearily stagger out of bed and look out the window. Most of them are at the far end of the field, munching on an early-morning bite of grass. There's still some pathetic-looking individuals standing next to the gate, heads through the bars, forlornly staring down the drive.

It's the same today – lots of bleating and shouting, but fortunately no escapees. Little Titchy seems one of the most unconcerned. He's lying under the tree in the sun, contentedly chewing his cud with all four legs tucked neatly beneath him.

Saturday, 18th August

It's the village fete today. I've had over twenty years in volunteering at these happy events, and done everything from managing a flatulent Santa who overindulged on parsnip soup to controlling a rabid crowd of toddlers at a glitter tattoo stall. This year I've volunteered to manage the hook-a-duck stall.

I offered to manage the hook-a-duck stall for 2018 on the actual day of last year's fete. I wanted to get in there first; otherwise I'd be left with some apocalyptic scenario such as running the ice cream stall or judging the flower-arranging competition.

The best thing about my stall is the fact that I don't have to move from one spot all day. I prepare carefully with two seats (one for me and one for any friends I might spot on the day), a table for my flask of tea, a gazebo to protect me from howling winds or scorching sun and a box of chocolatey prizes. Which I try very hard not to eat.

Dad has come to help. He sits in one spot, takes the money and makes loud comments about the behaviour of everyone's sticky and snot-encrusted children.

The village fete has a boat race, and Steve and the kids have made a sort of catamaran out of two stapled-together milk bottles, which is raced down the stream against a flotilla of other home-made boats. I turn into a madwoman, belting down the riverbank, bosoms revolving, pushing everyone out of the way while shrieking at our boat to go faster. We lose. To a last-minute entry consisting of an old fag packet with a lolly stick stuck to the top.

My kids beg for money, so I empty my purse of 10- and 20-pence pieces and pour them into their pockets. Otherwise they will spend all day standing two inches away from my ear droning 'Muuuuum. Muuuuuuuum. Muuuuuuuuuuum' in a flat monotone until I give them handfuls of more 20-pence pieces to spend on the tombola.

Tombolas are to children what crack cocaine is to a drug addict. Lucy and Ben spend £8.40 on tombola tickets in the first fifteen minutes before returning, proudly clutching a weathered Yardley toiletry set and a dusty bottle of pink bath salts.

The day is warm and dry. I have many tiny visitors who crowd around my plastic duck pond, and I am busily occupied handing out fishing lines and trawling for upside-down ducks. One child falls in, and lots of passing dogs help themselves to a drink.

At the end of the day I'm full of ice cream and warm white wine, and have a general haze of bonhomie towards the rest of our village. We troop back to the car carrying our various loot. I have bought five second-hand books and the kids are crowing over their prizes, which seem to be mostly sweets, grubby soft toys and the odd dog-eared comic.

Sunday, 19th August

The sunset this evening is stunning. All the hot weather has kicked up dust into the atmosphere and the colours are spectacular, from a deep crimson shading through peach to a delicate pink that makes the clouds look rosy against the deep-blue sky.

A very smart couple is sitting on my garden wall watching the display as I use a watering can to throw water on my flowers in the garden. They've been at a christening party at the brewery.

'Is it always like this?' asks the suited man.

'Most of the time,' I say, wondering if he means the sunset, or the weather in general.

'It's just so beautiful. I'd love to live here.'

I chat for a while and recommend that they go and stand at the gate into our barley field to watch the sunset. They set off, her in strappy high heels and tight dress, and him in a smart blue suit and brown brogues. Later on, I see them stood in the entrance to the ripening barley, her head on his shoulder, as they watch the last rays of sun peek across the ash trees at the bottom of the field.

Tuesday, 21st August

There's a farm sale advertised in the *Hexham Courant*. We love a good sale. This one is in north Northumberland, so Steve, the kids, Dad and I all pile in the car and take the hour drive up the Coquet Valley. It's being held at a farm right on the top of a hill, and the car parking is in a bumpy field next to the farm entrance. It's high and exposed up here, and when I open the door the wind almost rips the handle out of my hand.

The farmer is retiring, and all his machines and the contents of his barns are being auctioned by Hexham and Northern Marts. We don't know the family or the farm, but I can't help feeling sad that they are moving on, especially as they seem to have been here a long time, judging by the rows and rows of old tools and machinery.

The kids are having a great time, jumping from a pile of old tyres onto the top of a flat trailer. Dad is carefully sifting through a box of old tools and making a note of the ones that he wants to buy. I'm more interested in the food van, and haul everyone off for a bacon sandwich and a cup of piping-hot tea.

The auctioneer stands out in the field and moves methodically from lot to lot, selling everything to the highest bidder. He quickly sells a long line of old tools then moves on to a couple of vintage milk churns, and then finally the carts and trailers. Dad has bought an ancient-looking workshop vice and is happily cleaning it with his handkerchief. We bid on a seed drill, but the price goes too high so we all wander off to look at a beautifully refurbished grey Fergie tractor and an enormous modern combine harvester. There's a very big crowd, and we see a few faces we know.

It starts to blow a real gale but still the auctioneer carries on, wanting to get through all the outside lots before he can start on the cows and calves in the sheds. Everyone crowds round an old Fordson tractor and there's a brief but loud outburst of bidding before the new owner waves his card at the auctioneer, and then trails off to pay for his purchase in the auction's mobile office. Steve is deep in conversation with a local farmer, discussing the lamb prices and whether he should sell in the autumn or wait until the new year.

Even though we couldn't afford to buy anything we've had a good day, catching up with the farming gossip and eating a bacon butty for lunch. Dad falls asleep on the way home, snoring loudly while wedged between the kids in their car seats.

Wednesday, 22nd August

The wind is battering the roof of the house, wobbling the slates and booming down the chimney breast.

Last night the weather forecaster announced that the North East would feel the tail end of a big storm and it's definitely arrived in Northumberland.

I keep inside during the worst gusts, as the trees along the drive bend alarmingly, and shoals of sticks and leaves are dashed across the farmyard. We have a huge solitary ash tree in the back field that used to sit on a long-disappeared field boundary. It has an enormous trunk, pitted and twisted with many lumps and bumps, and must be at least a few hundred years old. In the wild weather the tree branches violently twist and sway, and I pray that it isn't felled by the storm.

The bad winds continue until after dark, and we sit listening to the sounds of twigs hitting our slates and the wind moaning around the windows and doors.

Thursday, 23rd August

It was one of the worst storms we've had all year. In the morning the road outside our house is littered with branches and twigs. Steve goes out to do an initial reconnaissance of the damage and comes back in with his shoulders slumped.

'There's three trees down,' he says, 'one in the barley, one in the hay field and one in the front field.'

When a tree falls in a crop field it damages the valuable harvest and prevents the combine harvester from cutting that area. We'll have to shift it before the crop ripens by sawing up the trunk and bigger branches and then collecting all the leftover wood and larger twigs. And we need to do it in the next three weeks before the oilseed rape and the barley is ripe and our harvesting swings into action.

'Right,' says Steve. 'I'll add it to the to-do list.' At least we'll now have enough firewood to last us for the next few years.

Friday, 24th August

We've been chopping and carting away the downed tree in the hay field all morning.

I've dragged the fat pony along for a bit of exercise and company.

I don't have a cart that she could pull, so instead I tack her up, tie a pair of seed bags together with baler twine and sling them over her saddle so that they hang down on either side. Now I can fill them full of the twigs from the trees.

She doesn't seem to mind, and after carefully turning around to investigate each bag in case it contains sheep nuts, she settles down to munch away at the field borders while I pick up twigs.

I always think picking twigs is a medieval sort of job, as it can't have changed in hundreds and hundreds of years. It's not difficult. You clear the grass fields of all the big branches and larger twigs before they're cut for hay, otherwise the downed wood is caught in the mower and your forage is full of chopped-up bits of wood.

Away from the downed tree the hay field is hip-deep in wonderful thick green grass dotted with yellow meadow vetchling, creamy-coloured meadowsweet and carpets of pink- and purple-tipped clover flowers. Earlier in the year I heard the plaintive 'cour-leee' of curlews and the high-pitched 'pee-whit' of lapwings that were nesting in the field, but now there's just a background hum from bumblebees and the sound of the grass itself, bending and rustling in the wind.

I trundle the wheelbarrow up to the tree, fill it full of cut logs and push it back to the farmyard, hauling fat pony and her bags of twigs along behind me. She occasionally jerks her head down to reach a particularly luscious clump of grass, and I stop for a moment, straighten my back and push on to dump our loads of sticks and logs in the corner of the barn. By the time we're finished the pony has eaten her body weight in grass and we have a respectable pile of cut ash logs, which will slowly dry out, ready for our log burner in the colder months.

Saturday, 25th August

We push our way into the thick stalks of the wheat and barley fields. The crops are almost ready, the golden stalks bending under the weight of the rapidly ripening seeds. The seed heads need to have less than 25 per cent moisture to be ready for harvesting. Steve has a clever little electronic moisture meter but prefers to check by biting a seed between his back teeth, even though he's at risk of cracking a molar. If the seed makes a crackling sound it's dry enough to cut.

If the crop is cut while it's too damp it has to be dried in a commercial grain dryer, which costs a lot of money.

The oilseed rape is also looking ripe, with the dark-yellow plant stalks turning a dark grey, and the little spiky pods rattling with the tiny black seeds.

The weather is set fair for the next few days, and Steve is constantly preoccupied, checking his phone for updates from John, the combine driver, and Don, the wagon driver who picks up our wheat, rape and barley harvest. It's a delicate business. We're only a tiddly farm in the big scheme of things, and don't have much clout, so we walk a thin line of checking that we're on their list of farms to visit and making sure that we don't harass their staff.

Of course, the decent weather means that every farm is doing the same thing, and the combine drivers are working flat out, right round the clock, trying to catch the sun and get the harvest in before the weather breaks. If it does rain during harvest we will have to stop and wait until the crop dries again, to make sure it's in optimum dry condition before cutting.

Steve isn't eating very much, is grumpy and stressed, and keeps sniffing the wind like Mavis. And if he can't get on with something (the ground is too wet, or the tractor needs fixing) he's almost in actual physical pain.

Tonight I look out the window and see lights dotted around in the distance right up to the horizon – all the neighbouring farmers drilling, ploughing, cutting, baling while the weather holds.

Sunday, 26th August

The combine arrives at 11 a.m. and steams into the yard. She's huge and bright yellow, with a strip of cab in the front shaded with darkened glass. John the driver waves at the kids and rumbles off to cut the first field. Steve rushes to his tractor, swings himself into the driver's seat and trundles off after him, pulling our long green trailer behind.

It's an impressive sight, the combine rattling along the field, with Steve sticking like glue to its side while the grain pours from the unloading arm and heaps into our trailer. They're gone all day, and when Steve comes back he's grinning from ear to ear, covered in a dark-brown dust. The barley crop is in, and we all troop out after supper to inspect the mounds of dark, golden seed heaped up in the shed.

That evening the porch is full of little piles of yellow barley seed as people tip out their boots or undo trouser turn-ups and scatter the barley across the stone floor.

Monday, 27th August

The combine is back to cut the wheat crop, and brewery visitors stand and stare while John expertly guides the massive monolith through the farmyard.

Steve eventually returns to the house late in the evening, chuckling to himself.

Tonight, when John had finished cutting the barley, he undid the header bar,* removed it from the front of the combine and hitched it up behind our tractor so Steve could drive it back behind the combine.

They were trundling in convoy along a single track, and

---

* A combine harvester has a detachable 'header bar' or cutting bar at the front, and it's so wide it doesn't fit along our narrow country roads so has to be unhitched, turned horizontally and drawn along behind with a separate tractor.

as they turned the corner they met a couple of ladies driving down the road in a tiny red car.

Apparently, the driver immediately stopped the car in a panic, and tried to back into a gateway. But she stalled her car horizontally across the road, and then sat frozen in her driving seat with her hands covering her eyes.

John and Steve waited for a bit, but she didn't take her hands away from her eyes or attempt to move the car.

After a while Steve got down from his tractor, gently helped her out of the car, reversed it into a gateway and then let the combine past.

'Maybe she thought if she couldn't see the combine, it wasn't really there. Did you get cross or sweary at them?' I asked.

'Nooooooo. Well, John rolled his eyes a bit, but I was very patient.'

I can sort of understand how the poor lady felt, as when I first saw a combine I couldn't believe how bloody enormous it was. When you spot one chuntering away in a field it doesn't look so big, but when it's bearing down on you in a narrow country lane, it's so massive it blocks out the light. And the full-length windscreen means you can see the driver staring beadily at you and gesticulating to get your tiny car out the way.

Wednesday, 29th August

Harvest is a stressful time of year. At least at lambing time Steve and I are both working together, going through the same problems and managing the farm. But at harvest time we only have one tractor, so I just see a distracted Steve occasionally, when he rushes back to ring a contractor or grab a cup of tea.

When I first married Steve I had visions of having harvest picnics in our stubble fields in lovely sunshine, with apple-cheeked children wearing tasteful Boden clothes, sitting on a

Cath Kidston rug, eating wholesome, home-made sausage rolls with lashings of ginger beer.

Instead, this evening, when I haul the kids out for a picnic in the stubble field, it's cold, the fields are prickly, the kids are arguing over who last went on the iPad and hate my home-made sandwiches, and all we've got to sit on are a couple of slimy sheep feed bags.

The kids spot Daddy in the next field, doggedly following the combine up and down in his tractor, and they wave their jumpers at him, trying to get his attention. He notices us, and waves back, the sun glinting off his windscreen. The combine is cutting the oilseed rape and there's a cloud of dust following them both, while a stream of black seed pours from the unloading arm into the back of Steve's trailer.

Thursday, 30th August

In our sheep shed we now have three 100-ton piles of wheat, oilseed rape and barley grain. The golden-and-black seed looks a little like sand, and it's with much difficulty that I persuade the kids that they can't jump into the mounds and scatter it about as if they were on the beach.

We'll store the piles until Don, the wagon driver, arrives with his articulated lorry to transport the grain to a holding depot a few miles away. We're part of a grain pool with a few other smaller farms, which gives us more clout when our harvest is sold to the big grain traders.

I spot Marjorie and Ethel determinedly marching towards the nearest pile of barley and we shoo them out of the barn, their wings flapping in fury. The last thing I want is the chickens eating the grain and depositing large amounts of chicken poo into the harvest.

Steve is proud. The land has averaged four tons of wheat to the acre, and three tons to the acre for the barley and the oilseed rape. These are very respectable figures, and will bring in a much-needed cheque from the grain buyers. The wheat goes for animal feed, the barley to be malted for beer and the oilseed rape is turned into cooking oil and bio diesel.

Friday, 31st August

Don has arrived, and Steve spends all day filling the lorry with huge scoops of grain.

We're glad he's turned up so soon after the harvest, as sometimes if there's a long wait the piles of grain start heating up, and then Steve is stressed and spends hours staring at the piles and moving the grain around with shovels in the hope that it doesn't get too hot.

We all help by brushing up any dropped seed, and I have a happy few minutes chatting to Don and listening to his reminiscences. He's planning to retire soon, which is a pity, as he's a first-rate driver, and manoeuvres the enormous articulated lorry into our farmyard with skill and panache.

Saturday, 1st September

My wonderful parents have bought a weekend's holiday in a cottage by the Northumbrian coast for the kids and me. (Steve can't come as he's still knee-deep in the harvest, cutting the last fields of wheat.)

I'm now sitting on a lovely beach with the sun belting down. Despite the heatwave I'm displaying such a bluish-white luminous skin tone it looks as if I've only just survived a terrible nuclear winter. Instead of a bikini I'm wearing a sensible floral swimming costume from Marks & Spencer, topped off by a straw hat with a chin strap. I realise halfway through the afternoon that I look just like Thora Hird.

Sunday, 2nd September

The kids are loving the sea, diving in and out of the waves like tiny porpoises. I've been persuaded to go swimming, and I venture a slow breast stroke before being slapped in the face by a long green strand of seaweed.

I retire to sit on a rock, book in hand, watching the kids jump the waves and gambol in the surf like a couple of bronzed puppies.

I miss Steve. I wish he could come with us, but he rarely gets time away from the farm, and when he does, he worries about the crops and animals left behind.

I'm not sure I'm built for holidays. The cottage is wonderful, and it's a very welcome break. But I must admit that I'm fretting about the animals left behind too, and whether Steve is managing on his own, and whether the harvest is finally finished.

Monday, 3rd September

Back at home I'm met by Heather holding an extremely grumpy-looking Cinders the kitten.

She's in disgrace. Heather found her on top of the remains of a hog roast, all four feet braced against the side of the turnspit while trying to peel off an overlooked strip of crackling.

'I'm so sorry!' I say, grabbing the cat, who hangs down either side of my grip like a deflated, stripy balloon.

I promise to keep her inside the next time there's a wedding, and shove the tiny cat back into the house. She kinks her tail, turns her back in disdain and starts to carefully lick her fur back into place.

Tuesday, 4th September

Harvest is over. Thank god.

Steve has lost half a stone in stress and worry, but the weather has held, and we now have a barn full of sweet-smelling hay bales, and all the fields have been cut.

We're debating whether to bale the straw or just chop it and plough it back into the soil.

There's a UK-wide shortage of bedding and forage due to the previous cold winter, so we would gain a significant income from selling straw bales to local farmers. However, this is balanced against the need to add good fertiliser into the soil. Chopped straw is free, and is a great way of nourishing the crop fields without buying in an expensive top dressing such as compost.

'Why can't you just spread the stuff in our manure pile instead?' I ask naively.

'Because it'll only fertilise about ten acres, and then I'll have to buy it in for the rest of the fields,' he says.

He finally decides to bale the majority as straw, as at least then we can sell it to local farmers who are struggling for bedding, and chop the smallest field to plough into the soil.

I take a walk with Mavis along the golden, dusty stubble fields. She plunges into the grass at the side of the field, searching out rabbits. There's a family of pheasants scratching about in the soil and they scatter as she bounces through them, squawking in indignation as they fly into the air.

Wednesday, 5th September

The hedges are groaning with blackberries, so the kids and I take as many pans and bowls as we can and start picking the fruit. Mum and Dad come to help, and there's a companionable silence as we work along the hedge together, punctuated by the occasional yelp as a thorn penetrates a sleeve or trouser leg.

Mum wanders off to the bottom of the field. Suddenly there's a shout and a splash as she puts a foot straight into the stream that girdles the stubble. Apparently, she was stretching over the water to reach a luscious-looking bramble when the bank gave way and she ended up in the water.

She squelches back to the house while we totter behind with pans crammed with fruit and our mouths stained purple with juice. I towel her off and we sit in front of the TV, picking through the fruit and eating any squashed berries.

Our hedges always produce masses of fruit, so I'll make blackberry pie, crumble, jam and eventually, in desperation, freeze as much of the brambles as I can.

Thursday, 6th September

We're weighing all the lambs to make sure they're over 40 kg, and heavy enough to sell at the Mart. Steve herds them into the shed through a narrow sheep race and then pushes them one by one into a small pen containing a flat metal weighing plate.

As they step on the plate the scale calculates their weight, and it's satisfying to see that every single lamb is 40 kg, and some of them are much heavier.

The two 'Spotty Brothers' are now mammoth, fully grown sheep, with thick legs and broad snub-nosed faces. Fred and Fuzzy come up to mid-thigh, and have the typical 'Beltex' look with a wide, flat face and exaggeratedly muscled backsides. Even Titchy reaches the 40 kg mark, and is now a round, fat woolly oval with a chubby tail.

The lambs bounce and kick through the race, jinking from side to side and digging in their feet, refusing to stand on the unfamiliar scale. Steve and I sweat and pant, hauling them from the front and pushing from behind. They bolt back into the pens and blare and jostle each other while ramming their bodies against the gate, where Mavis is watching with narrowed eyes. Cute they are not.

Friday, 7th September

I'm staring out the window when a bevy of lads wearing luminous yellow donkey jackets appear outside and start unloading wheelbarrows, shovels and a pile of black road chippings. It's Northumberland's Pothole Repair Crew. Our roads are appalling, and after the bad winter and dry summer the potholes are increasing. They have already burst a couple of tyres and jiggered the suspension on passing cars. I ring round the neighbours in excitement to report the news. I stand in the kitchen window, cup of tea in one hand, and watch the crew start to fill in the holes.

There's a young lad with a brush that cleans up the pothole, a man with a spade who shovels in the road chippings and another with a big roller that squashes them all down. Everyone looks exceedingly grumpy.

Every time a car comes up the road they all down tools, laboriously move the wheelbarrow and the roadroller to the side and stand on the verge as the car drives past. Usually the driver winds down the window and says something along the lines of 'About time!', 'Bloody good job you're doing there!' or 'Make sure it lasts through the winter'.

One of the men then gets all red about the ears and obviously wants to welly the drivers with his spade.

I'm on tenterhooks to see if he starts to shout at people. Eventually they move on to the next batch of potholes up the road. I reckon there's enough work to keep them going right through the summer, autumn and into next January.

Saturday, 8th September

I have spotted seventeen lapwings in our back field. I know that we had a pair of curlews nesting in the hay fields but I'd no idea there were so many lapwing babies.

I've been hearing them call since early summer. It's a sort of 'wooble wooble woowoo weeeeweee wip wip' sound.

They have little black-and-white bodies with a tuft on top. I try to take a photo to post on Twitter, but due to the poor quality of my camera phone they come out looking like blurry sparrows.

Monday, 10th September

We've run all our lambs into the sheep sheds to sort through them in preparation for selling them at the Mart tomorrow.

They need dirty bottoms clipped out and their feet dressing to make them look presentable.

It's a hard and heavy day, but worthwhile if we can make the lambs look good enough to squeeze a few more pounds out of the buyers.

Out of the 290 lambs, around twenty ewe lambs are kept back for breeding, to replace the older ewes in our flock.

They look fabulous. Randy and Thrusty have done a good job passing on their Beltex muscling, and the lambs look fit, fat and stocky. We leave them in the paddock overnight and spend the evening creating sheep-movement forms, checking the catalogue and writing in the Mart pen cards.

Tuesday, 11th September

It's a very busy day at the Mart, and we're drawn seventh on the sale ballot. We wait our turn, looking over our pens of lambs, scratching a neck here and there as they chew their cud and itch against the bars.

The whole family goes into the ring, Steve shooing the lambs in front while the kids and I bring up the rear. As the bidding starts I try and catch the eye of the buyer, imploring him to bid to a good price, while the children stare with big eyes at the auctioneer and try to catch what he's saying.

Eventually, our pens are sold for an average of £75 per lamb.

This is a great price and, rejoicing, we make our way back to the ring, congratulating each other on a good sale.

The kids are given a pound each to go and present the buyers with 'luck money'. This is a custom I found astonishing when I first met Steve.

'You mean you give your money to the people that have just bought our lambs?!' I said in amazement.

'Yes, a couple of quid here and there. Just means that next time there's a sale they might remember us and bid a little higher.'

Most of the buyers (god rest their souls) refuse the luck money from the kids, and instead tell them to spend it on themselves. One cheery buyer gives the children a fiver to split between them, and they rush off jubilantly to spend it on sweets and pop at the café.

This is the culmination of a year's worth of toil, sweat and tears. It only takes around two minutes to sell each pen of lambs, and the small profit from each animal will be ploughed back into the farm, to pay our debts, to buy more animal feed and crop seed, and to maybe buy some new animals to extend our flock a little.

We march back to the car and empty trailer, chatting excitedly about the good day we've had and calculating how much money the Mart cheque will bring.

Wednesday, 12th September

Today the farm seems quiet without our flock of bolshy, noisy lambs. The ewes are grazing peacefully in the fields, and the stubble has a bevy of partridge pootling around pecking at the leftover seed.

We're back at the beginning of our farming year – and it's almost time to get Thrusty and Randy into the sheep pens again, to give them the once-over before tupping season starts.

It's been a year of ups and downs – the weather, the lack

of money and the sheer hard work of hanging on to a farm by our fingernails has been the overriding theme. But, more importantly, we're still here. And I love our little patchwork of fields and woods and buildings, and if we can manage to keep our heads above water for the next few years, to keep High House in good nick to pass on to Lucy and Ben, I'll be incredibly grateful.

I walk towards the sheep and Mabel totters over for a neck scratch. I bury my face in her thick wool and inhale her wonderful sheepy smell. She gently huffs at my face, searches briefly for some sheep nuts, then ambles back over the short-cropped grass, to return to her flock.

A Note on the Type

Diary of Pint-Sized Farmer has been set in Elena, a typeface designed by Nicole Dotin in 2007. The strokes of the typeface bear the influence of a broad-nibbed pen, yet the type manages to feel formal without being stuffy.

Book design adapted by Brooke Koven

1970–2020
David R. Godine
Publisher
FIFTY YEARS